P9-BUH-150

BLUEPRINT FOR LIFE

JOURNEY THROUGH THE MIND AND BODY

Other Publications:
WEIGHT WATCHERS® SMART CHOICE RECIPE COLLECTION
TRUE CRIME
THE AMERICAN INDIANS
THE ART OF WOODWORKING
LOST CIVILIZATIONS
ECHOES OF GLORY
THE NEW FACE OF WAR
HOW THINGS WORK
WINGS OF WAR
CREATIVE EVERYDAY COOKING
COLLECTOR'S LIBRARY OF THE UNKNOWN
CLASSICS OF WORLD WAR II
TIME-LIFE LIBRARY OF CURIOUS AND UNUSUAL FACTS
AMERICAN COUNTRY
VOYAGE THROUGH THE UNIVERSE
THE THIRD REICH
THE TIME-LIFE GARDENER'S GUIDE
MYSTERIES OF THE UNKNOWN
TIME FRAME
FIX IT YOURSELF
FITNESS, HEALTH & NUTRITION
SUCCESSFUL PARENTING
HEALTHY HOME COOKING
UNDERSTANDING COMPUTERS
LIBRARY OF NATIONS
THE ENCHANTED WORLD
THE KODAK LIBRARY OF CREATIVE PHOTOGRAPHY
GREAT MEALS IN MINUTES
THE CIVIL WAR
PLANET EARTH
COLLECTOR'S LIBRARY OF THE CIVIL WAR
THE EPIC OF FLIGHT
THE GOOD COOK
WORLD WAR II
HOME REPAIR AND IMPROVEMENT
THE OLD WEST

For information on and a full description of any of the Time-Life Books series listed above, please call 1-800-621-7026 *or write:*
Reader Information
Time-Life Customer Service
P.O. Box C-32068
Richmond, Virginia 23261-2068

BLUEPRINT FOR LIFE

JOURNEY THROUGH THE MIND AND BODY

BY THE EDITORS OF TIME-LIFE BOOKS
ALEXANDRIA, VIRGINIA

CONSULTANTS:

KEN BROWN, who studies the role of neurohormones and stress proteins in the regulation of gene expression, teaches in the Department of Biological Sciences at George Washington University, Washington, D.C.

WESLEY BROWN teaches at the University of Michigan, Ann Arbor; his research there is in the area of evolutionary genetics and molecular biology.

SEAN CARROLL teaches molecular biology and genetics at the University of Wisconsin, Madison.

PAULA GREGORY is Education Director of the National Institutes of Health's Human Genome Center in the Department of Internal Medicine at the University of Michigan, Ann Arbor.

MICHAEL KASTAN is a pediatric oncologist and geneticist at the Johns Hopkins University, Baltimore.

JEANNE M. MECK is director of the Cytogenetics Laboratory at Georgetown University Medical Center, Washington, D.C., where she teaches in the Department of Obstetrics and Gynecology.

REED PYERITZ, clinical director of the Center for Medical Genetics at the Johns Hopkins University, Baltimore, researches the diagnosis and management of hereditary disorders.

NANCY SEGAL is the director of the Twin Studies Center at California State University, Fullerton. She also teaches there in the Department of Psychology.

WAYNE STANLEY is director of cytogenetics at the Children's National Medical Center, Washington, D.C., and teaches at George Washington University, Washington, D.C.

IRWIN WALDMAN teaches in the Department of Psychology at Emory University, Atlanta, where his research interest is developmental psychopathology.

NELSON WIVEL is director of the Office of Recombinant DNA Activities, National Institutes of Health, Bethesda, Maryland.

JOURNEY THROUGH THE MIND AND BODY

Library of Congress Cataloging-in-Publication Data

Blueprint for life / by the editors of Time-Life Books.
p. cm. — (Journey through the mind and body)
Includes bibliographical references and index.
ISBN 0-7835-1004-7 (trade) —
ISBN 0-7835-1005-5 (library)
1. Human genetics—Popular works.
2. Heredity, Human—Popular works.
3. Genes—Popular works. 4. Nature and nurture—Popular works.
I. Time-Life Books. II. Series.
QH431.B622 1993
573.2'1—dc20 93-21603

First printing. Printed in U.S.A.
Published simultaneously in Canada.
School and library distribution by Silver Burdett Company, Morristown, New Jersey 07960

This volume is one of a series that explores the fascinating inner universe of the human mind and body.

CONTENTS

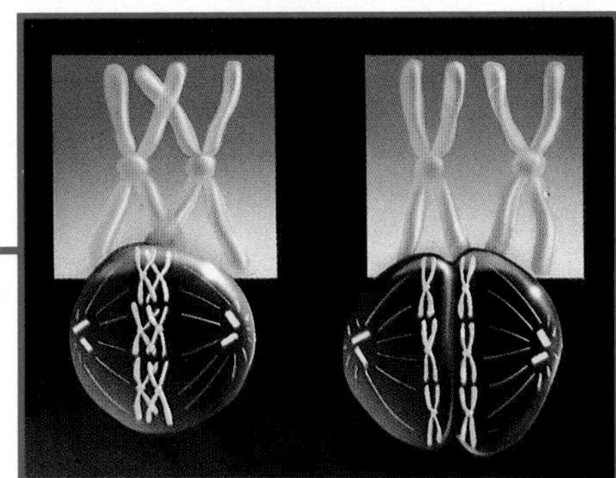

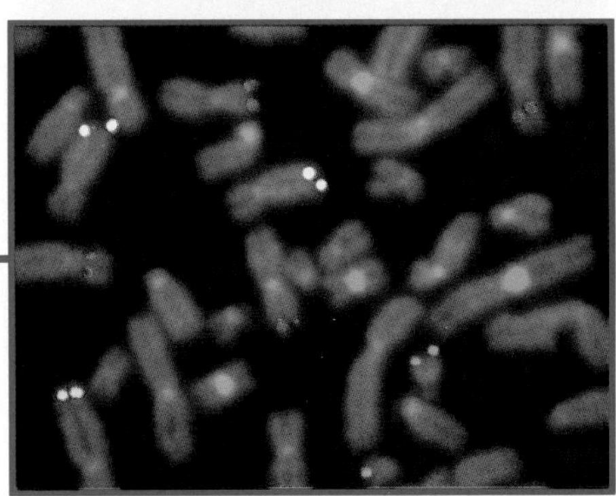

22
CAPTAIN
3

1

The Miracle of Inheritance

Even astounding coincidences generally hold only passing interest for serious scientists, who see them as mere quirks of probability. At first, the story of Jim Springer and Jim Lewis, identical twins who had been separated a few weeks after birth, seemed to be a case in point. When the 39-year-old brothers were reunited in 1979 and began comparing notes, they discovered a series of unusual parallels in the details of their lives. In addition to being given the same first name by their adoptive parents, each had an adoptive brother Larry and as boys had had a dog called Toy; they had worked at McDonald's and at gas stations and had also served as deputy sheriffs. Both had been married twice, first to women named Linda, then to women named Betty; Jim Lewis had a son named James Alan, Jim Springer a son named James Allan. Both routinely vacationed along the same stretch of beach in Florida; drove the same color, make, and model of car; drank the same kind of beer; and chain-smoked the same brand of cigarettes. And both lived in towns about equidistant from their birthplace in Piqua, Ohio, in the only house on the block with a white bench around a tree in the yard.

These remarkable similarities attracted a good deal of media attention when they came to light, leading to

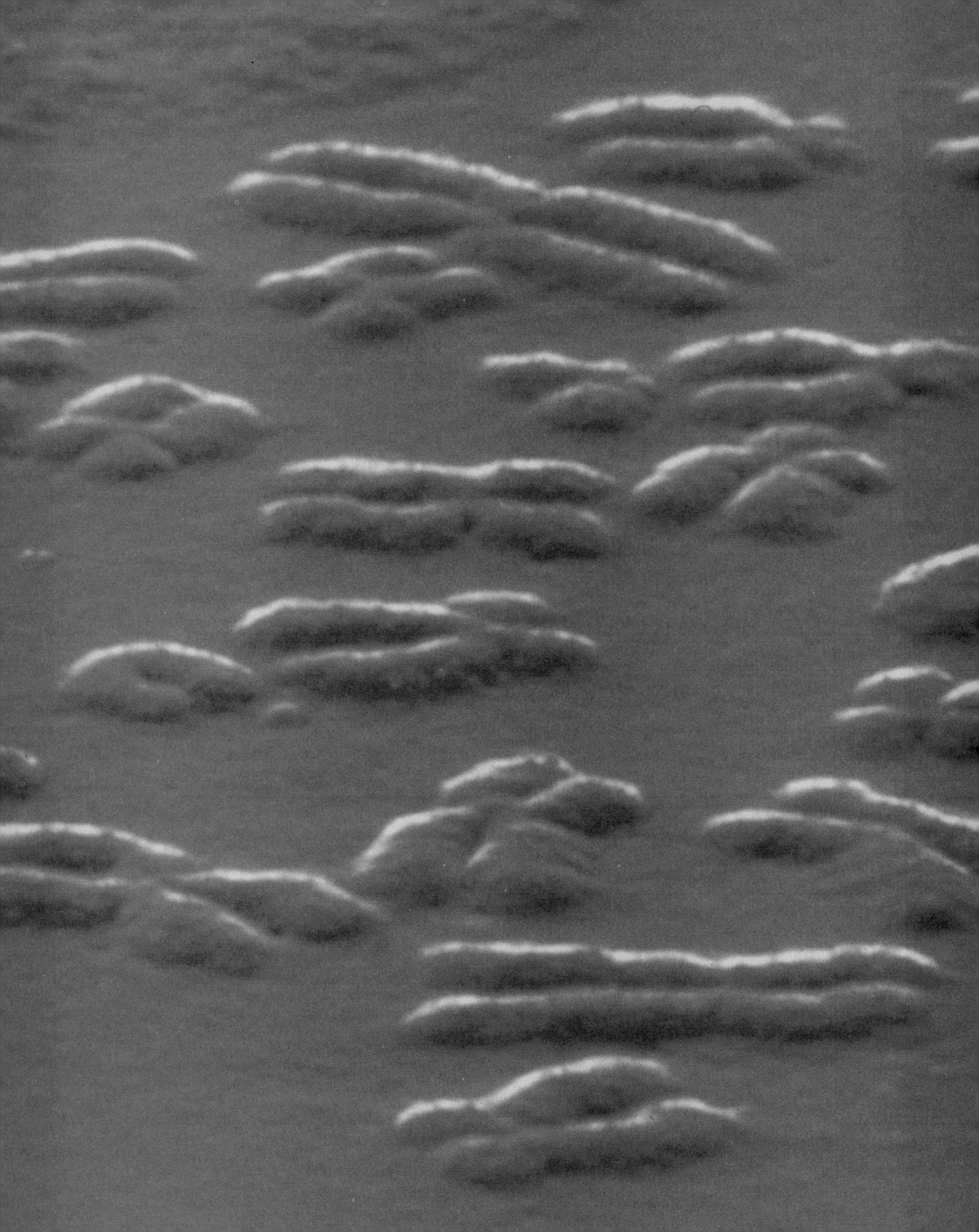

PRECEDING PAGE: A cluster of human chromosomes from a white blood cell appear tens of thousands of times life size in this false-color image taken through an electron microscope. Almost all cells in the body contain 23 pairs of chromosomes, packed with all the genetic information necessary to produce a complete human being.

feature articles on the so-called Jim twins in several national magazines and to appearances on television talk shows. But in scientific circles, the two long-separated brothers were potentially more intriguing for how they might be different than for how they were alike. University of Minnesota psychologist Thomas J. Bouchard, Jr., who had been studying environmental influences on behavior, was especially interested: Twins such as Lewis and Springer were identical in terms of their inherited biology, so any differences between them would have to stem from environmental factors. Within an hour of first reading about them in a local paper, Bouchard had arranged for a grant that would allow him to study the twins in a clinical setting, and just over a month after the two were reunited, they were beginning the first of a series of interviews and tests at Bouchard's laboratory. As Jim Springer later noted, "Dr. Bouchard said he wanted to get us before we were—what was the word he used?—'contaminated' by being around each other too much."

In a sense, however, the contamination appeared to be inherent. The more Bouchard learned about the twins, the more uncanny resemblances he found between them. To begin with, in addition to the obvious likenesses in appearance, the two Jims had startlingly similar medical histories. Both had high blood pressure, had experienced severe migraine headaches beginning in their teens, and had suffered what they thought was a heart attack at one time. Physiological parallels were one thing, but what really surprised Bouchard and his colleagues were the twins' mental and psychological profiles. Their scores on all sorts of personality evaluations—rating such attributes as sociability, self-control, and open-mindedness—were virtually indistinguishable, mirroring those that might be produced by a single individual tested on two separate occasions. Standard intelligence measurements also matched closely. Perhaps the most objective gauges were electroencephalograph (EEG) readings that recorded the electrical activity of their brains in response to various stimuli: The recorded patterns looked so much alike that, once again, the researchers were hard-pressed to tell Jim Springer and Jim Lewis apart.

As convincing as anything were what Bouchard called "the little things that go together to form a personality"—from posture and body language to facial expressions and subtle intonations of voice. As far as these elements were concerned, the psychologist commented, the two Jims "were like bookends."

Bouchard was stunned by the results. They seemed to fly in the face of the accepted wisdom that the features of a person's character—including general disposition, likes and dislikes, attitudes and beliefs—are more a product of upbringing and environment than of heredity. With the typical restraint of a scientist, Bouchard stopped short of drawing definitive conclusions from his observations of Springer and Lewis. In fact, in an effort to quantify as best he could the balance between nature and nurture, he would go on to conduct an extensive study of more than a hundred other sets of twins who had also been raised apart (Chapter 3). For the time being, however, he was willing to take one tentative but significant step down a road he had not expected to travel: The case of the Jim twins seemed to suggest that, as Bouchard put it, "the genetic effect pervades the entire structure of personality."

That so-called genetic effect traces to one of the most wondrous aspects of life on Earth. Coiled within nearly every cell of the human body—and, for that matter, within the cells of every

Despite having been raised apart, identical twins Jim Lewis *(top)* and Jim Springer *(bottom)* have much more in common than just physical traits: The chain-smoking brothers spend many hours in their basement workshops on carpentry projects, and both especially enjoy making miniature furniture.

type of living thing from bacteria to whales—is a remarkable blueprint, a complex set of coded instructions written in molecules and handed down from generation to generation, in some instances over billions of years. This blueprint plays an essential role in the life of every organism, steering its growth and directing its day-to-day cellular activity. Known technically as the genome, it also determines what makes each species unique. And in myriad ways, it shapes the uniqueness of each individual within a given species.

The physical components of that individuality are clearly apparent in our own species. The human genome encompasses a wide assortment of varying physical characteristics—from such obvious ones as hair, eye, and skin color to the more subtle details of facial features and fingerprint patterns. With the notable exception of twins, triplets, and the like, these elements have come together to form as many different combinations as there have ever been people.

But as Bouchard and others have found, the blueprint extends even further, into such realms as intellect and temperament. To what degree it does so remains the subject of intense debate, with opinions covering a wide spectrum. At one extreme are those who argue that all aspects of human behavior can be traced back

to the inherited biological program; at the other are those who regard heredity as essentially insignificant in the formation of personality—running a distant second to environment even in its effect on physical development. Although most scholars hold to the middle ground, agreeing that nature and nurture interact in many ways to shape an individual, the findings of 20th-century research seem to have tipped the scales somewhat toward biology. At the very least, the information in the blueprint is now acknowledged to exert a much broader influence than was once thought.

Less than 150 years ago, virtually nothing was known about the mechanisms of inheritance. Since then, the science of heredity—or genetics, as it came to be called (from the Greek word *genes*, meaning "born")—has advanced by leaps and bounds, and at a breakneck pace in the last 50 years. By and large, the progress has been a matter of shifts in focus down to more and more fundamental levels of detail, starting with studies of how physical traits pass from one generation to the next, and ending with examinations of the molecular alphabet in which those traits are encoded. Indeed, by 1990 investigators were well under way with a concerted effort to lay bare—letter by letter—the entire set of genetic instructions that go into making a human being.

The knowledge explosion has revolutionized genetics itself. No longer content simply to observe and theorize about the workings of life's blueprint, researchers now take an active role in changing it. Of course, human beings have been manipulating nature for millennia, ever since prehistoric societies first began to crossbreed plants to encourage desirable traits. But today's experimenters, armed with sophisticated laboratory tools and a wealth of detailed genetic information, are able to tinker directly with the blueprint's instructions, introducing into the cellular machinery alterations that would never have arisen naturally and that would have been impossible to achieve through crossbreeding.

So far, most of the genetic manipulation has been carried out with bacteria, crops, domesticated stock, and laboratory animals. But scientists have already had limited success in treating human ailments by dosing patients with genetically altered cells, attempting to correct genetic flaws at their cellular source. The efforts have raised ethical concerns among philosophers, theologians, and policymakers, and in the scientific community as well: After all, the same methods used to combat disease might eventually be employed to select "ideal" characteristics for an offspring—to engineer nature in ways that many people find deeply disturbing.

Attempts to fine-tune the genetic code point to one of the hallmarks of heredity: its ability to combine constancy with variation. Students of the earliest forms of life on Earth have shown this feature to be an evolutionary development. The first single-celled organisms that emerged on the fledgling planet more than three billion years ago were programmed only to preserve an original set of attributes. They reproduced by simple fission, splitting apart and passing on exact copies of the coded instructions that directed all their biological activity. Bombarded by intense solar radiation through the planet's thin, just-forming atmosphere, and subjected to all sorts of chemical compounds in the soupy oceans of the young Earth, these cells would often undergo mutations that changed their programming. Then, through natural selection, the versions of an organism that were best suited to their environment would thrive—and forward these new traits to succeeding generations.

Somewhere along the line, something happened—scientists are still not sure what—that altered the mechanism of reproduction itself. At least

TEEMING VARIETY. The 100,000 or so genes that constitute the human blueprint come in so many versions and so many different combinations that no two people—with the possible exception of identical twins—have ever looked exactly alike. Such extreme individuality is a characteristic of the human species.

570 million years ago by most estimates, certain organisms developed the ability to reproduce sexually, that is, by uniting with another member of the species rather than by merely splitting apart. Instead of passing on an exact copy of their biological code, the organisms conveyed a shuffled set of instructions to their offspring, half from one individual, half from the other. This new form of reproduction ensured that the next generation, while maintaining most of its species's attributes, would be different in some way from the previous one. Variation thus became an integral part of the process rather than just a chance occurrence, and the stage was set for a tremendous burgeoning not only in the diversity of species but also, in some instances, in the diversity of individuals within a species.

In the case of humans, of course, it is news to no one that children can be both like and unlike their parents, that a son can have his mother's eyes and his father's nose, or be so much a mixed bag of characteristics as to resemble neither one of them all that much. Such vague aspects of inheritance had long been known, but as late as the middle of the 19th century, not even the most learned biologists understood the rules that dictate how specific traits and combinations of traits are passed on. Why, for example, did certain attributes seem to skip a generation, disappearing in the children only to reappear in the grandchildren?

The first great insight into those rules, and the first inkling of the mechanism behind them, came from the work of an Austrian monk by the name of Gregor Mendel, who had a keen interest in both botany and general biology, stirred no doubt by a childhood spent on his father's farm and orchard. Shortly after entering the Augustinian monastery in Brünn (now the Czech city of Brno) in 1843, Mendel satisfied his scientific curiosity as best he could by rummaging through the monastery's library, which contained a number of books on horticulture and agriculture. Mathematics also appealed to him, and he served for a while as substitute teacher in the subject at a local secondary school. But by 1856 his inquisitiveness was at work in a less scholarly setting: the monastery garden, where he began a series of experiments that are universally acknowledged as having founded the modern science of genetics.

The target of Mendel's investigations was the common garden pea. He had observed a clear-cut pairing of alternate characteristics in seven distinct categories. A plant would have, for example, either tall or short stems, its pods would be either green or yellow, or its peas would be either smooth or wrinkled. He also found that, when the plants were left to pollinate themselves, some bred true to form, so that a given characteristic such as tallness appeared in every subsequent generation. Mendel then decided to see what would happen when true-breeding plants with one characteristic were crossbred with true-breeding plants having the alternate characteristic. The process involved meticulous hand pollination of hundreds of plants and an equally meticulous charting of results through several generations of offspring.

One of Mendel's experiments involved fertilizing tall-stemmed plants with pollen from short-stemmed ones and then planting the resulting seeds.

He discovered that the next generation of plants all had tall stems, the short-stemmed trait having apparently disappeared completely. But when Mendel allowed these plants to self-pollinate (or cross-pollinated them with each other), he was surprised to find that they yielded both tall- and short-stemmed plants, in a ratio of about three to one. Furthermore, when these plants in turn were allowed to self-pollinate, the short-stemmed ones bred true to form, as did a third of the tall ones, but the other two-thirds of the tall plants produced offspring with the same three-to-one mix of tall and short stems.

Mendel found the same patterns of trait inheritance for the seven categories he had identified. The results led him to two significant conclusions. First, one of the alternatives in each pair of traits clearly had the power to override the other, as could be seen when a cross of tall and short plants initially produced only tall versions. He called the stronger trait dominant and the weaker one recessive. But how could a recessive trait reappear in later generations? The key was Mendel's brilliant deduction that offspring inherited from each parent one of a pair of what he called factors, units of hereditary information each associated with a given trait.

According to Mendel's theory, differing combinations of dominant- and recessive-trait factors accounted for the varying proportions of characteristics he observed in different generations. To help him keep track, Mendel adopted a convention of labeling the factors for dominant traits with capital letters and recessive alternatives with lowercase letters, a practice still followed today; a true-breeding line for a dominant trait would be labeled AA, for example, and the recessive version aa. To trace the patterns, he began from the understanding—aided by his knowledge of statistical laws—that inheriting one of two factors from each parent meant that there were four possible combinations of factors in offspring. In the first crosses he made between true-breeding lines, the four combinations were identical: Aa in every case, because the offspring could inherit only a dominant-trait factor from one line and only a recessive-trait factor from the other. And only one trait—the dominant one—showed itself.

But then things got more complicated. When he crossed these hybrid Aa plants with each other, he realized that the four combinations of factors would come in three forms—AA, Aa, and aa—and that the Aa form would occur twice as often as the other two,

A FAMILY TREE FOR PEAS. By crossing different varieties of pea plants, Gregor Mendel was able to deduce that a combination of two factors—later known as genes—determines whether a given trait is dominant or recessive. The diagram below illustrates the pairing of factors resulting from a cross between yellow-seeded (AA) peas and a green-seeded (aa) strain, with each subsequent generation having been allowed to self-pollinate. Both the AA and the Aa combinations of factors give rise to yellow seeds, but only the aa pairing yields green ones. The yellow trait is thus the dominant one.

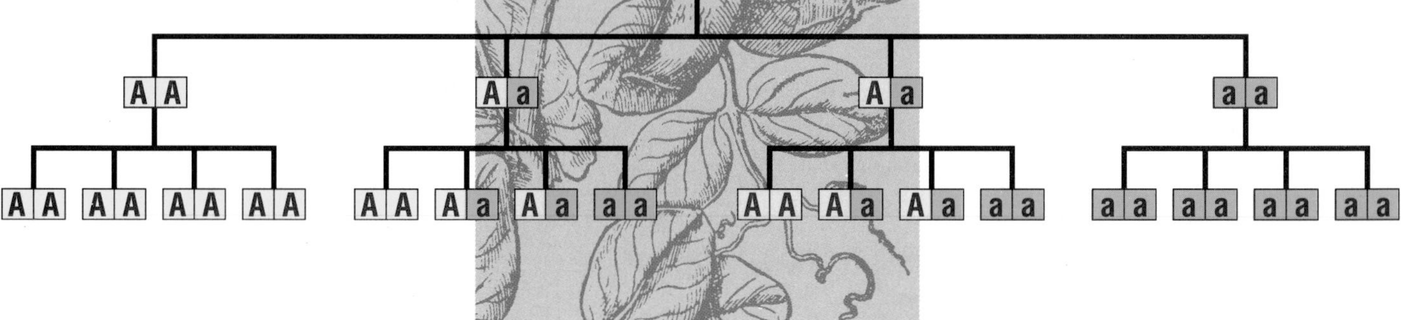

since either factor could be inherited from either parent. The theory coincided with his experimental results of a three-to-one ratio of traits because both AA and Aa, accounting for three-quarters of the offspring, produced the dominant trait, while only aa—occurring only a quarter of the time—yielded the recessive trait. But it was not until the next generation of plants that the theory proved out. The dominant-trait line that bred true and arose from a third of the dominant-trait plants could only have come from AA parents, which his statistics had demonstrated would be present in precisely a third of the cases.

Mendel presented his results in papers read at two 1865 meetings of Brünn's modest Natural Science Society, which published them the next year. They were masterpieces of logic and mathematical acumen, but he had been able to take the story only so far. Where he could not go was down to the cellular level to divine where and what his so-called factors were, and how they exerted their indisputable influence.

For this and other reasons, his work never reached the wider scientific community in his lifetime. Mendel became the abbot of his monastery three years after his initial reports were published, and he no longer had time to pursue the research; his papers languished in obscurity for 34

Redefining Mendel's Rules

One of the central tenets of Gregor Mendel's rules of inheritance is that a given gene has precisely the same effects on offspring regardless of whether it is inherited maternally or paternally. But this and other aspects of heredity have come under fire in the light of recent discoveries. Indeed, some researchers speculate that genetics is on the verge of a revolution in thinking as profound as that introduced by Mendel more than a century ago.

Among the first investigations to challenge the accepted wisdom was a 1989 study of two rare genetic diseases. Researchers found that children suffering from a condition known as Angelman syndrome—characterized by mental retardation, uncontrolled laughter, and stiff, puppetlike gestures—had a flaw in a particular segment of one chromosome. Another disorder, Prader-Willi syndrome—which causes a milder form of retardation and has different physical manifestations, including extreme obesity—turned out to result from precisely the same flaw on precisely the same chromosome. The only difference was that Angelman sufferers had inherited their defective chromosome from their father, and Prader-Willi victims from their mother.

This odd disparity may be an evolutionary development stemming from a conflict of interest between the sexes at the genetic level. Experiments with mouse embryos have shown that paternal copies of certain genes tend to promote fetal growth, while the maternal versions restrict it to some extent. The theory goes that by doing all they can to improve the embryo's chances of survival, the paternal genes help ensure their own perpetuation. But maternal genes look out for themselves in a different way, by limiting fetal growth and thereby protecting the mother's health so that she will be able to produce more offspring later—and thus more copies of the maternal genome.

LIKE BEGETS LIKE. The unmistakable resemblance of a father and son provides clear evidence of the passing of genetic information from one generation to the next. Although the mechanisms of heredity ensure a certain amount of variation between parent and offspring, the blueprint for many basic features—such as the shape of an ear or the profile of a face—often remains unchanged.

years before being rediscovered in 1900. Meanwhile, techniques involving the use of microscopes had advanced to the point that scientists were indeed able to peer into living cells, discern their makeup, and begin to unravel some of the mysterious processes at work within them.

As far as the study of heredity is concerned, the most important discovery emerged in the 1870s. Biologists had already determined that most cells in the body contain a nucleus and that many also divide in the process known as mitosis, forming two new cells. To see more clearly what was going on in the nucleus during mitosis, they applied different staining dyes and then observed that just before division took place, a swarm of threadlike structures became visible within the nucleus. Because of the structures' affinity for dyes, they were called chromosomes, from Greek roots meaning "color body."

Later investigations revealed that chromosomes somehow have the ability to form copies of themselves, and that just before a cell divides the two copies split apart, with one half going to each of the two new cells. Thus, each new cell after division contains as many chromosomes as the original cell. There is, however, one notable exception. A special type of cell that became known as a germ cell goes through a different version of cell division called meiosis (*pages* 23-26), which results in the formation of reproductive cells—such as the sperm and ovum in humans—containing only half as many total chromosomes as the original germ cell. (This means, of course, that when reproductive cells unite during fertilization, the newly formed cell gets a full complement of chromosomes, half from one parent and half from the other.) The key to meiosis, researchers found, is that in one stage of the process, chromosomes organize themselves into like-size pairs, and when a germ cell splits, one member of each pair goes to each of the new cells.

Well before the turn of the 20th century, these findings were leading scholars to speculate that chromosomes were the carriers of heredity. Then, with the rediscovery of Mendel's rules of inheritance in 1900, the conclusion became inescapable. After all, chromosomes came in pairs just as did Mendel's "factors." Two scientists working independently, Walter Sutton in the United States and Theodor Boveri in Germany, are credited with the postulation in 1903 of what became known as the chromosome theory of inheritance, which spells out the direct link between the pairing of factors Mendel had observed and the behavior of chromosomes.

But there were still many puzzles. For one thing, in most species that were examined, there appeared to be many more of Mendel's factors than there were chromosomes. Researchers began to suspect that these factors, renamed "genes" by Danish scientist Wilhelm Johannsen in 1909, had to be subunits of the chromosomes. If that was the case, then some of the traits passed on from parent to offspring should travel together, because the genes that specified for them would be on the same chromosome. As logical as this seemed, it contradicted one of Mendel's central tenets, that traits were entirely independent of one another. He had never found in any of his experiments that the presence of one trait was linked to the presence of any other.

This turned out to be one of the few mistakes in Mendel's conclusions, one that stemmed not from an error in reasoning but from a strange quirk of circumstance. As researchers eventually discovered, Mendel's peas had only seven pairs of chromosomes, and the seven categories of paired factors he examined just happened

1 **THE NUCLEUS.** Often called the cell's information center, the nucleus holds all the genes that make up the chromosomes. The dark orb inside the nucleus is the nucleolus, which figures in the creation of ribosomes *(4).*

2 **CHROMATIN.** These compact, ropelike fibers contain DNA, which consists of genes. There are 46 such fibers in the nucleus, each of which contracts into a chromosome just before cell division.

3 **MESSENGER RNA (mRNA).** Derived from DNA, mRNA serves as the instructions for the making of a protein outside the nucleus. Some 10,000 varieties of mRNA may be found in a cell, one for each protein the cell produces.

4 **RIBOSOMES.** A cell may contain as many as 10 million of these tiny structures, which translate mRNA into proteins. Some ribosomes float freely in the cytoplasm *(12);* others sit on the rough endoplasmic reticulum *(5).*

5 **ROUGH ENDOPLASMIC RETICULUM (RER).** An extension of the nucleus wall, this mazelike element is named for its surface texture, made granular by ribosomes that produce proteins usually destined for use outside the cell.

6 **GOLGI APPARATUS.** This group of five to eight thin, overlapping structures accepts proteins produced at the nearby RER. After additional processing, the proteins leave the Golgi apparatus ready for work.

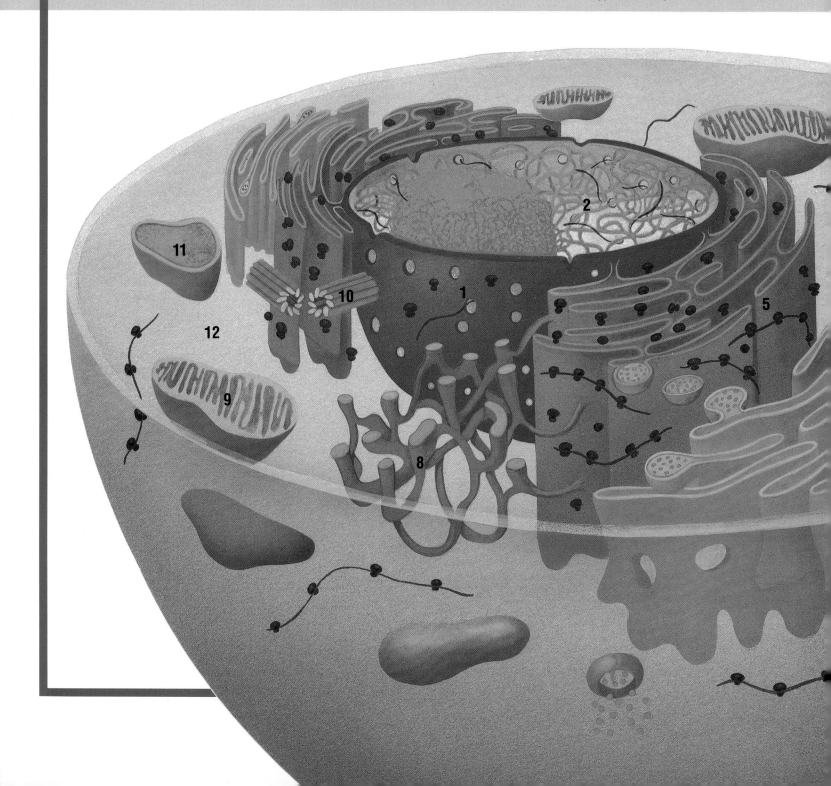

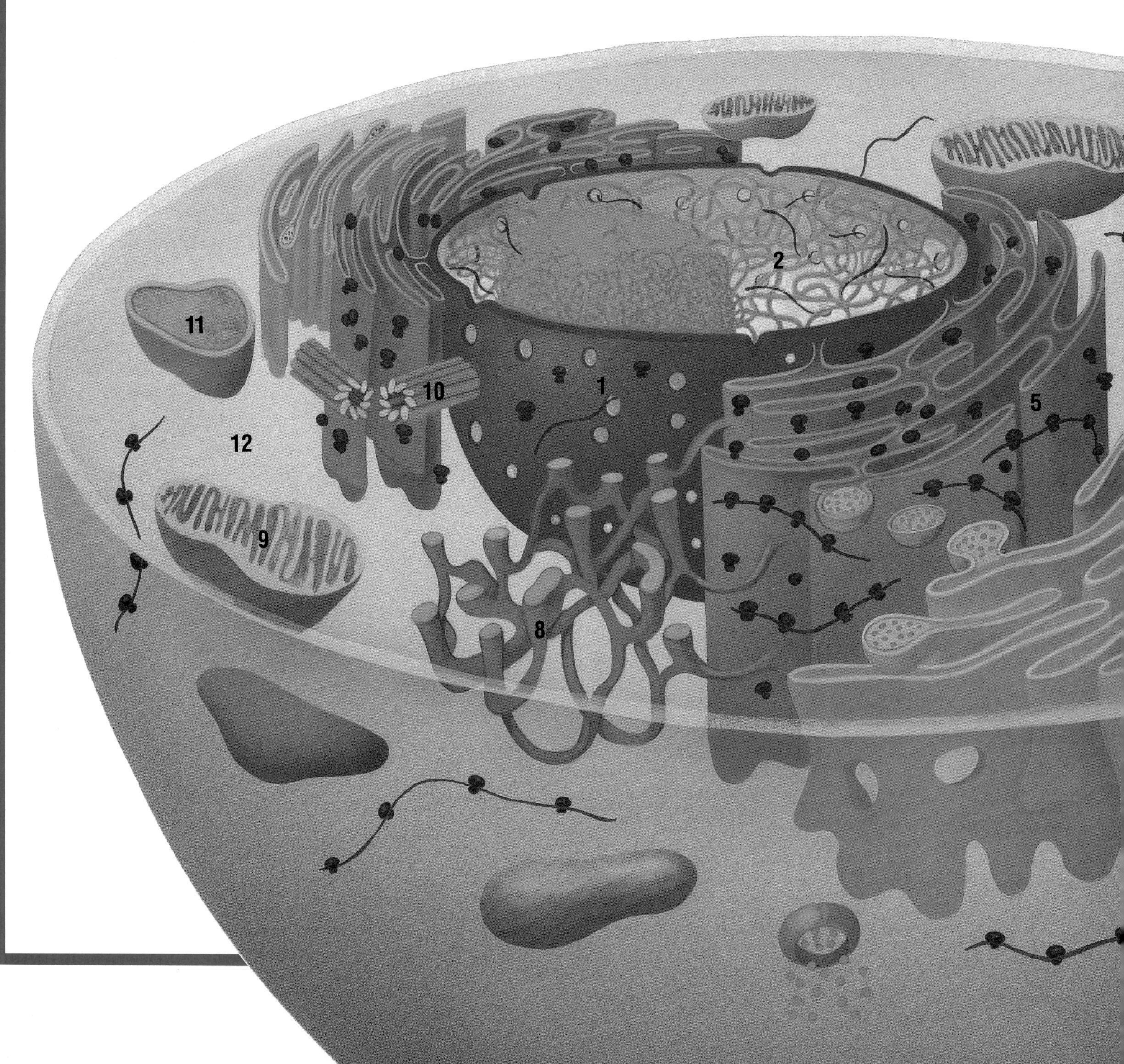

7 VESICLES. These pods carry protein molecules through the cytoplasm *(12)* to the Golgi apparatus and from there to the cell membrane, shielding the proteins from reactions with substances in the cytoplasm.

8 SMOOTH ENDOPLASMIC RETICULUM (SER). This tubular network, an extension of the RER but without ribosomes, plays various roles depending on the cell, from making cholesterol to neutralizing harmful chemicals.

9 MITOCHONDRIA. Fats and sugars give up their energy in these large structures, which store it in the chemical ATP. This compound is the power source for many cell functions, from protein synthesis to muscle-cell contraction.

10 CENTRIOLES. These two small cylinders each contain 27 tiny tubular elements. Centrioles double in number during cell division, when they help assure that each daughter cell gets a full complement of chromosomes.

11 LYSOSOMES. In the cytoplasm, lysosomes carry enzymes that digest sugars and fats and break down foreign substances. Lysosomes isolate the enzymes to prevent them from upsetting the cell's chemical balance.

12 CYTOPLASM. Occupying the space between the nucleus and a cell's outer membrane, cytoplasm consists of a viscous liquid called cytosol and structures such as the Golgi apparatus, lysosomes, and mitochondria.

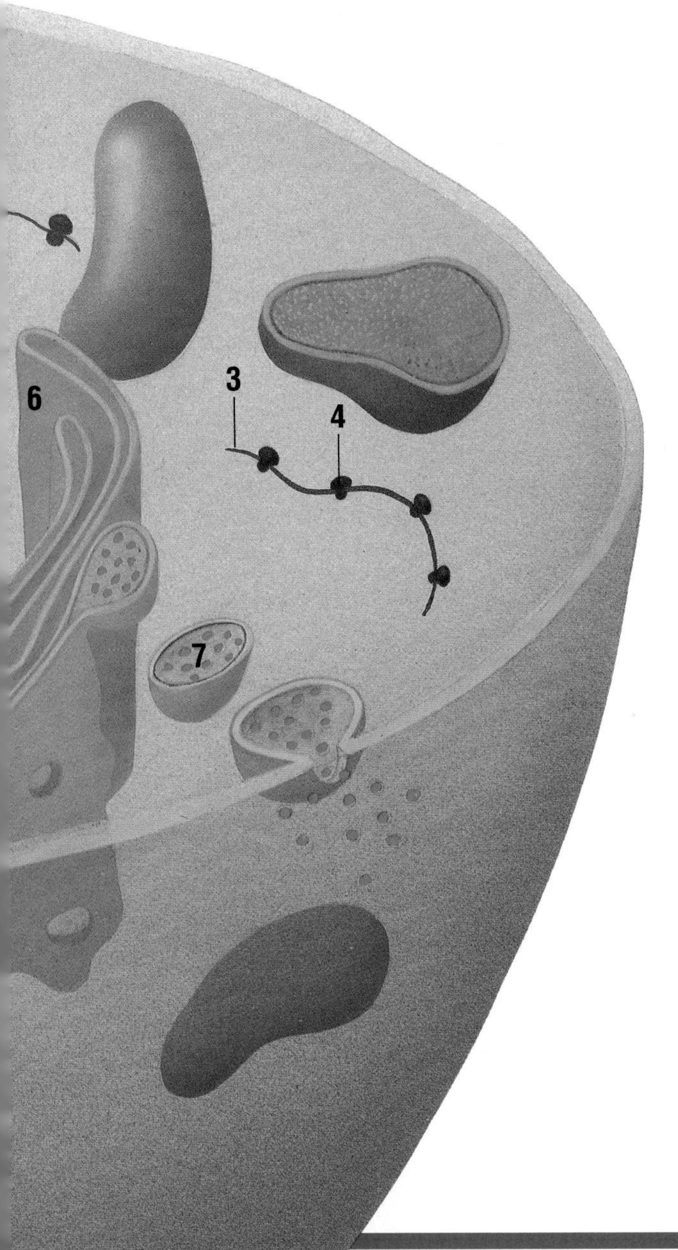

Guide to a Cell's Inner Works

Bone, skin, muscle—indeed all of the body's many tissues—are constructed of cells, the smallest units of human life. These building blocks are tiny—some 10,000 average-size cells from, say, the liver will fit comfortably on the head of a pin—and they come in an astonishing variety of forms and functions. Operating individually, cells carry oxygen from the lungs to every part of the body or devour bacteria and other harmful invaders to forestall disease. Acting in concert, they accomplish feats as remarkable as making the heart beat and storing memories in the brain.

Cells buzz with activity. Like any living organism, they take in nutrients and convert them to energy for a number of purposes. Muscle cells, for example, expend considerable energy to move an arm or a leg. Most cells also reproduce themselves and are called upon constantly to synthesize protein molecules, some for their own use and others for export. And all these ventures produce wastes, which must be disposed of.

The internal biological machinery for these processes, shown in its essentials at left, is shared by most cells. The single exception is the red blood cell. Produced by the bone marrow and supplied with a lifetime stock of oxygen-carrying hemoglobin, it has neither a nucleus nor any of the other components that employ the genetic material stored there.

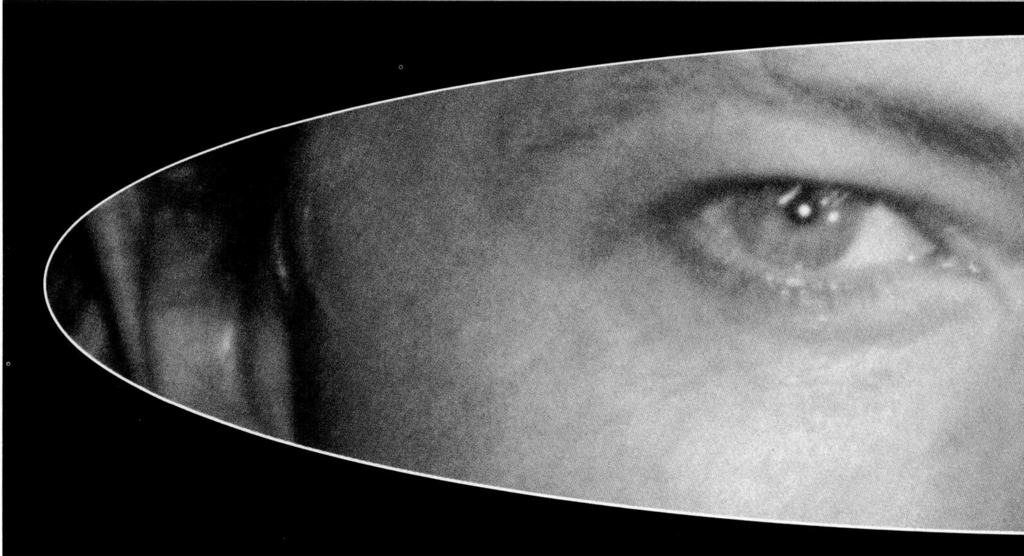

to be neatly dispersed among them. (In two cases, genes were actually on the same chromosome but, for complex reasons, were passed on as if they were on different ones.)

Mendel had been right in so many regards, however, that a challenge to the principle of trait independence could not rely on logic alone. But by 1910, American biologist Thomas Morgan had made a series of discoveries that finally settled the issue and at the same time led to one of the first definitive theories about the physical nature of genes. His test subject, among the most famous in the annals of genetic research, was the fruit fly *Drosophila melanogaster*.

Fruit flies had become a popular means of studying patterns of inheritance, because they produce new generations in rapid succession—about once every 10 days—and they also possess an assortment of traits, from eye and body color to wing shape, that are easily tracked. Morgan had been working for some time breeding multiple generations in his laboratory at Columbia University in New York City when he noticed one male fly with a trait he had never seen before: white eyes, instead of the normal red. Realizing that he had come upon a rarity, apparently the result of a spontaneous mutation of some sort, Morgan was extremely careful with the one-of-a-kind specimen and was even rumored to have taken it home with him at night in a small milk jar. He decided to mate it with a normal red-eyed female and then see what happened with the offspring.

He noticed right away that the white-eyed trait was recessive, because the first generation of offspring all had red eyes. But the results in succeeding generations revealed a series of new patterns, the two most significant of which were that the white-eyed offspring were much more frequently male than female, and the male offspring of white-eyed females were always white-eyed, which was more suggestive of a dominant than a recessive trait.

There was only one way to explain these findings, and it involved another momentous discovery that had been made several years earlier. Around the turn of the century, the American zoologist Clarence McClung had found that the cells of male and female grasshoppers differed in one key regard: In female cells, he could see that the two chromosomes in every chromosome pair looked exactly like each other; but in male cells, a certain pair consisted of one large chromosome and one small one. He called the large one X and the small one Y, and went on to theorize that this pair contained the information that determined a grasshopper's sex. Since each reproductive cell would contain only one member of this pair (an X or a Y in sperm, but always an X in an ovum), he rightly concluded that an embryo's sex was determined at the moment of fertilization, when two reproductive cells joined: The combining of an X and a Y made a male, while two Xs made a female.

Aware of McClung's discovery, Morgan was able to make sense of his fruit-fly results. If the gene for the recessive white-eyed trait was on the X chromosome, many more males than females would be white-eyed, because the males had no second X

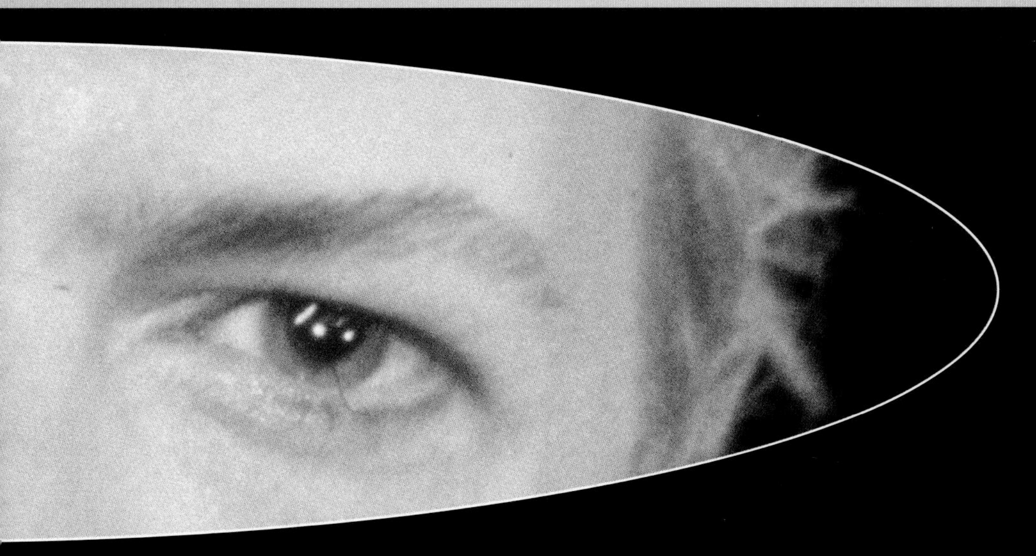

A MIXED GENETIC MESSAGE. Rock star David Bowie's unusual combination of eye color probably results from a mutation in a single gene. Although the pair of genes responsible for iris pigment normally affects both eyes equally, a mutation in one copy can lead to an uneven expression of the genetic trait, causing an absence of pigment—and thus a lighter cast—in one eye.

that might carry an overriding gene for the dominant red-eyed trait. Furthermore, a white-eyed female, both of whose X chromosomes had to carry the white-eyed gene, produced males with only white eyes, because the males' lone X chromosome could come only from her. In other words, this recessive trait acted like a dominant trait in males, because only the one gene on the X chromosome was necessary for its expression.

All this was interesting, but what was especially significant was that Morgan had conclusively demonstrated a link between a genetic trait and a specific chromosome—the first time such a connection had ever been proved. Then, in later experiments, he found other fruit-fly traits linked to the X chromosome, such as miniature wings and a yellow as opposed to a brown body. And with that, Mendel's rule of absolute trait independence became a thing of the past.

But Morgan was not done yet. Other results led him to realize that chromosome pairs occasionally exchanged segments—and the genetic information they contained—with each other. Through a sophisticated analysis of how this so-called crossing over took place (*page* 23), Morgan was able to draw up a map of the relative positions of dozens of genes on fruit-fly chromosomes. It was breakthrough work, published by Morgan and several colleagues in 1915 in *The Mechanism of Mendelian Heredity*. At its heart was a notion that soon came to seem obvious, but that had been in doubt before: Genes, the carriers of hereditary information, were physical entities distributed along chromosomes.

Morgan's work inspired a number of researchers, some of whom began to focus more intently on genes to discover precisely what they are made of and how they carry out the most crucial task of all: duplicating themselves in order to pass on their hereditary information. But there was equal fascination in using the newfound knowledge about genetics at a more macrocosmic level—to adjust and extend Mendel's original rules and to examine all their ramifications, particularly in the sphere of human heredity. Indeed, much was gleaned without any reference to the physical composition of genes, by analyzing their ultimate manifestation—inherited traits.

For example, not long after Mendel's papers resurfaced, an English physician named Archibald Garrod discovered that a rare disease he had been studying followed the rules for recessive traits as described by Mendel. It was the first of thousands of instances of human traits that have been shown to hew to Mendelian patterns. Garrod also noted that the disease could result from a defect in an enzyme, a type of protein that plays a role in such metabolic activities as digestion; he called it an "inborn error of metabolism." It was the first suggestion that the role of certain genes is to direct the production of substances that enable an organism to function properly.

As studies continued, researchers adopted a few standard terms that were, essentially, elaborations on Mendel's original designations. To begin with, they formalized the distinction between a gene itself and the different versions it could take, which became known as alleles (from a Greek root meaning "other"). Pea plants, for example, have a gene for stem height, whose two alternatives, or alleles, are tall and short. Also, an offspring that receives identical alleles from its parents—AA or aa in Mendel's terms—came to be called homozygous, a compound word deriving from the Greek *homos*, meaning "same," and *zygon*, meaning "pair"; one that inherits different versions of the same gene was said to be heterozygous (from *heteros*, meaning "different"). Although these terms can be somewhat cryptic to the uninitiated, they serve as an effective shorthand for referring to genetic possibilities.

The most famous Mendelian designation remained unchanged, however. A trait was said to be either dominant or recessive depending on how it passed from parents to offspring. But soon other wrinkles in the hereditary pattern came to light. Investigators found, for example, that a given gene can have three or more alleles, so that more than two possibilities exist for the resulting trait, and other designations besides dominant and recessive must be applied.

A familiar instance of this occurs in human red blood cells. With a few rare exceptions, the world's population shares only four main blood types: A, B, O, and AB. Around the turn of the 20th century, the Austrian-born pathologist Karl Landsteiner arrived at these categories by noting the presence or absence of certain protein compounds, dubbed A and B, on the outer coat of red blood cells. Later research revealed that there are three alleles involved: one for producing the A compound, one for the B, and one—designated O—that yields no protein compound at all.

A child who inherits two O alleles produces no compound and is classified as type O. A child receiving two alleles for A, or one for A and one for O, will produce the A compound and so will be blood type A; similarly, blood type B arises from a combination of two Bs or a B and an O. In these cases, Mendel's rules apply as usual, with both the A and the B alleles proving to be dominant to the O allele. But a fourth possibility exists. When one parent contributes an A allele and one a B, the child's red blood cells will have both compounds and the blood type will be AB. The A and the B alleles thus fall into a subset of dominant inheritance called codominance: Neither one overrides the other, and both manage to produce their related trait.

Sometimes the distinction between dominant and recessive blurs in other ways, as has been demonstrated with certain flowers such as the snapdragon. When a white-blossomed plant is crossed with a red-blossomed one, all the offspring in the next generation have pink or rose-colored flowers. As with the alleles for the two blood proteins, neither the white nor the red allele produces a dominant trait capable of overriding the alternative. The difference in this case is that in combination they yield an intermediate trait.

The picture gets even more complicated. The simplistic notion that every gene is responsible for a specific trait, and that every trait arises from a specific gene, soon fell by the wayside. In fruit flies, for example, researchers discovered that the gene affecting eye color also helps determine body length. Conversely, such traits as eye color in humans are actually the product of several different

Dividing to Unite: The Story of Sex Cells

When a sperm and an egg come together to form an embryo, they are completing the final step of a process that ensures the shuffling and reshuffling of genetic information from one generation to the next. Central to that process is the creation of sex cells in the first place, by a special kind of cell division known as meiosis, from a Greek word meaning "to reduce." Most human cells contain 23 pairs of chromosomes, with one member of each pair deriving from the mother and the other from the father. Normally when a cell divides, the two newly formed cells retain the full complement of chromosomes. But so-called germ cells, which develop into sperm cells and eggs, divide in such a way as to cut that number in half.

As illustrated here and on the following pages, division by meiosis involves two main phases and several key steps. In terms of the shuffling of genetic instructions, the most crucial events take place during the first phase, when chromosomes pair up. To begin with, maternal and paternal chromosomes organize themselves randomly in each pair, so that each newly formed cell will receive a mix of paternal and maternal chromosomes. And in a process known as crossing over (*below*), paired chromosomes actually exchange portions with each other, ensuring that the combination of genes on a particular chromosome will also vary.

The end result is a collection of sex cells each containing half the genetic information necessary to create a new life. The stage is thus set for reproduction—a literal reunion of chromosomes, newly arranged to produce a unique human being. (For simplicity, the 46 chromosomes in a human cell are represented in the illustrations that follow by three paternal chromosomes, shown in blue, and three maternal, in pink.)

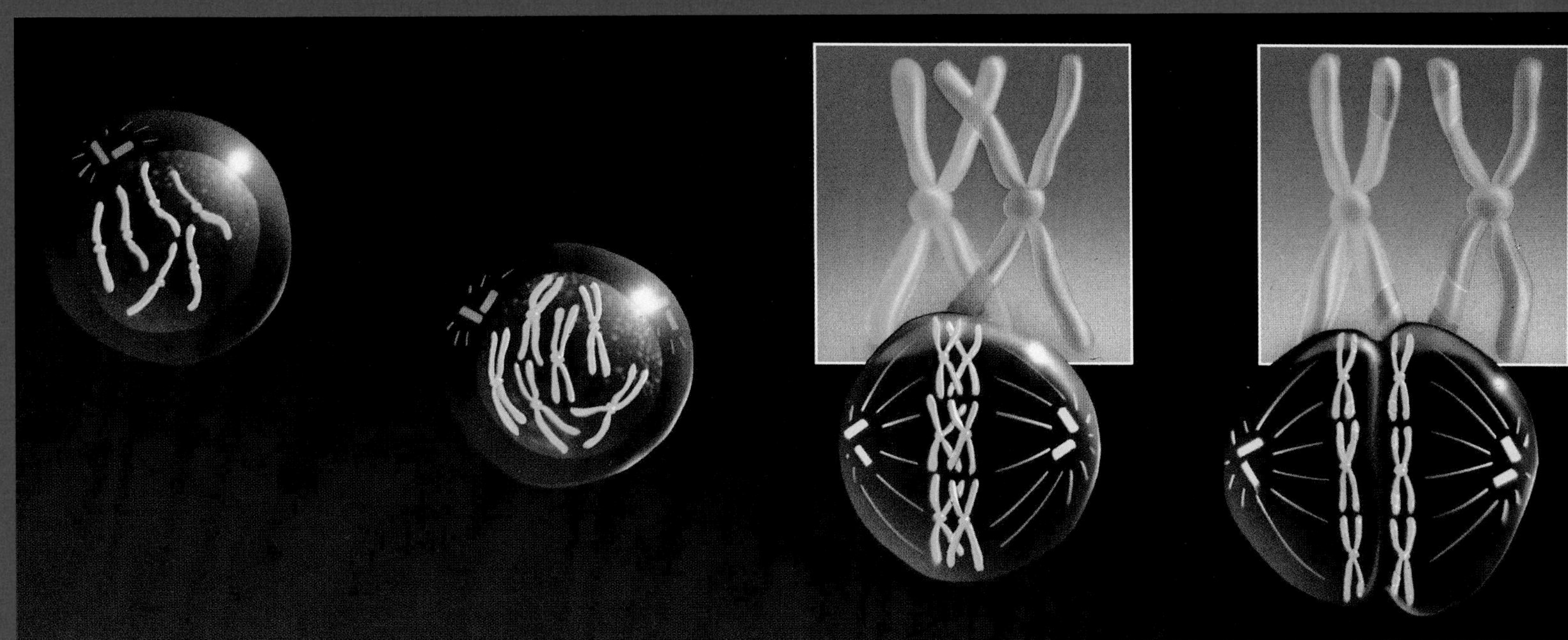

46 SINGLE STRANDS
Before meiosis begins, a germ cell's chromosomes consist of single strands scattered within the nucleus. Two centrioles lie outside the nucleus.

46 DOUBLE STRANDS
Each chromosome replicates itself to form a double strand; the centrioles also double and start to move toward opposite poles of the cell.

46 DOUBLE STRANDS
Chromosomes pair up, and fibers radiating from the centrioles cause them to align. In crossing over *(top)*, contact enables pairs to exchange segments.

23 DOUBLE STRANDS
The cell's outer surface has visibly begun to divide as the fibers draw chromosome pairs apart. Each chromosome includes maternal and paternal genes *(top)*.

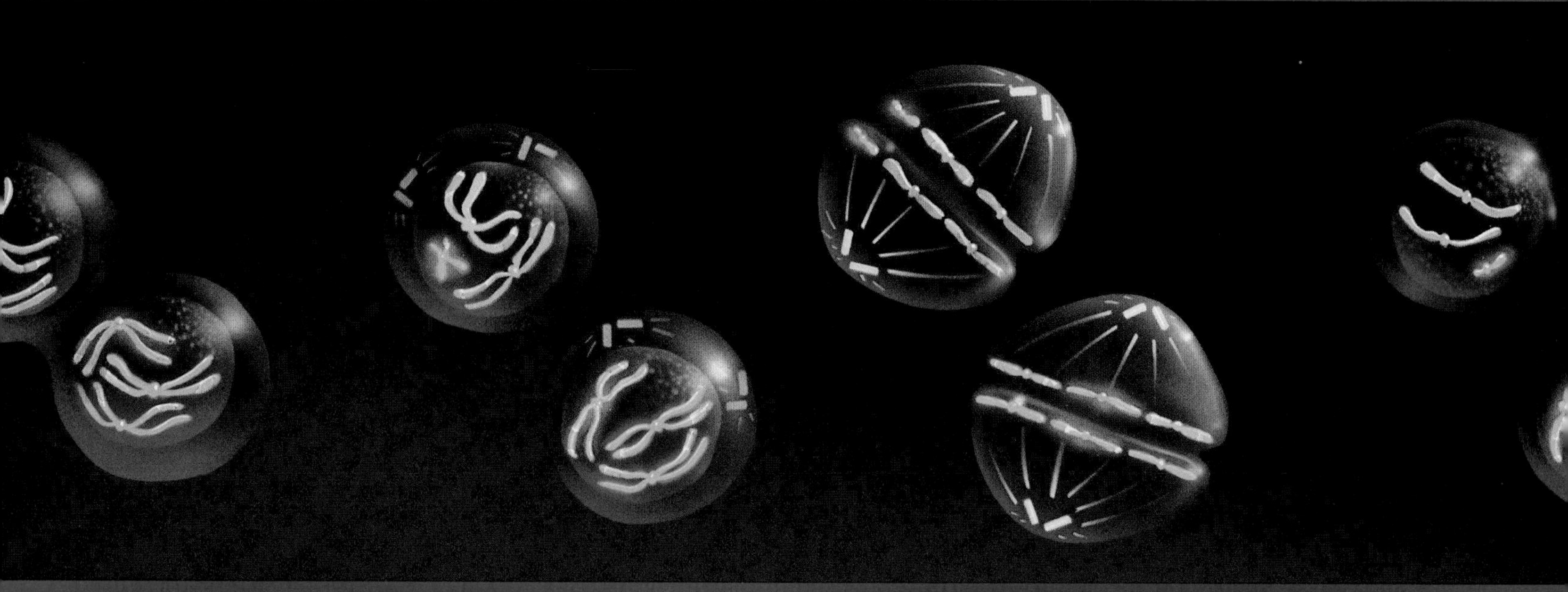

23 DOUBLE STRANDS
In males, division creates two equal-size cells, each with half the original set of chromosomes. At this stage, the outer cell membranes are still attached.

23 DOUBLE STRANDS
Each new cell has one of the two sex chromosomes *(highlighted):* the smaller Y chromosome *(top),* and the larger X. New centrioles have formed as phase two begins.

23 SINGLE STRANDS
Centriole fibers have pulled all the double strands apart. Each chromosome now includes its original amount of genetic information, instead of double.

23 DOUBLE STRANDS
In females, the division of cellular material is unequal, creating one larger and one smaller cell. However, both contain the same half-complement of chromosomes.

23 DOUBLE STRANDS
The larger cell will eventually give rise to an ovum, or egg. Both it and the smaller one, known as a polar body, carry an X sex chromosome *(highlighted).*

23 SINGLE STRANDS
A second division begins as outer membranes split and double-stranded chromosomes are pulled apart by centriole fibers. The larger cell divides unevenly.

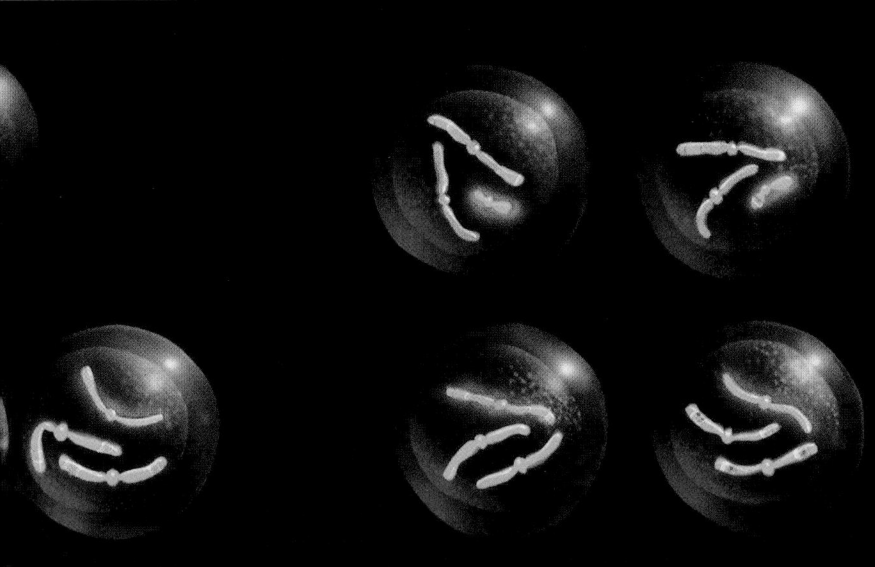

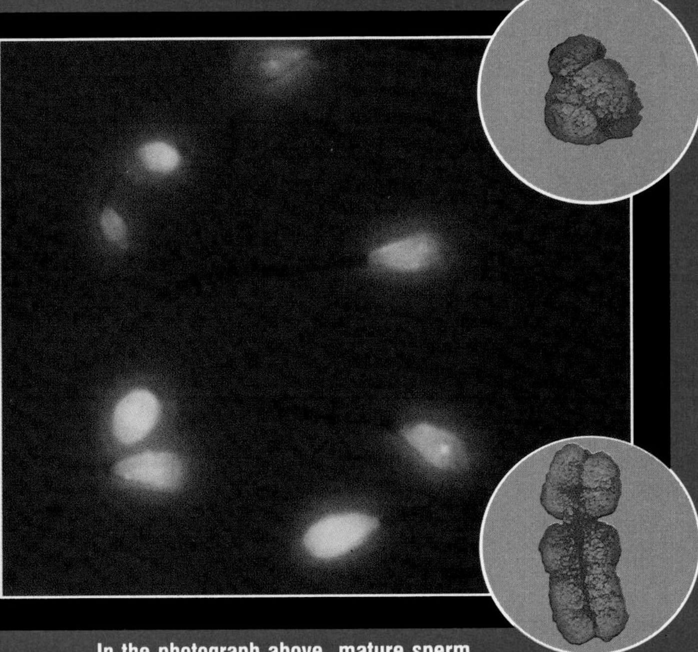

23 SINGLE STRANDS
As the second phase of meiosis nears completion, outer membranes of dividing cells remain attached. A nucleus has formed in each of the four new cells.

23 SINGLE STRANDS
The four new cells are now completely independent. Two contain Y chromosomes, and two contain Xs. Within about 16 days, all four will mature into sperm cells.

In the photograph above, mature sperm cells swim in a liquid specially treated to cause those containing a Y chromosome to glow bright yellow. The insets show closeups of a Y chromosome *(top)* and an X chromosome *(bottom)*.

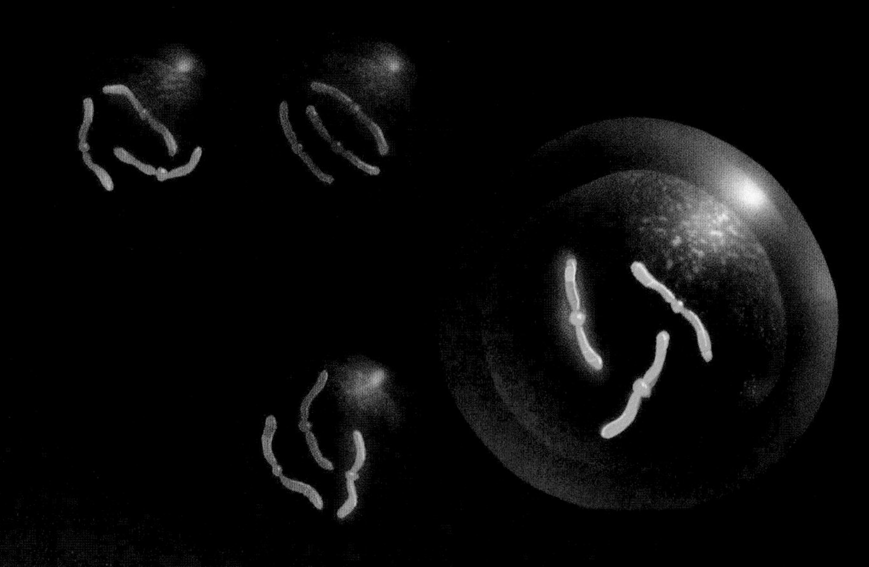

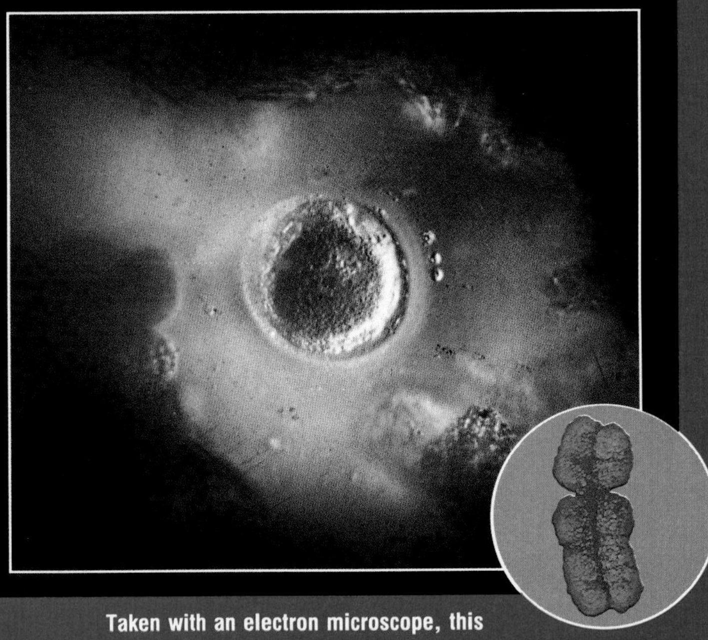

23 SINGLE STRANDS
The second phase of meiosis creates one large cell that will become an egg, and three polar bodies. The polar bodies will soon disintegrate.

Taken with an electron microscope, this image shows a mature human egg, ready for fertilization. The three small bubbles to the right are nourishing cells. Eggs contain only the X chromosome *(inset, right)*.

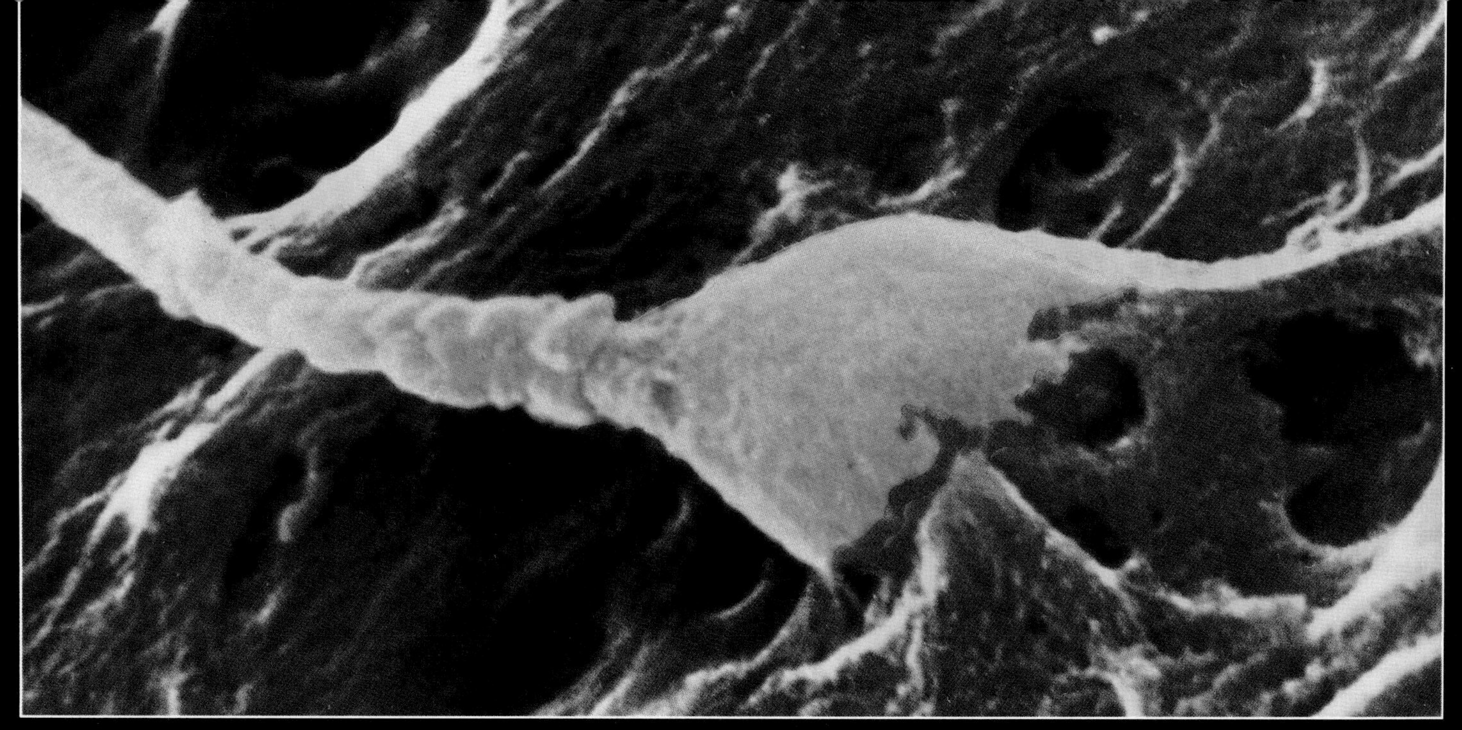

The Ultimate Union

The false-color image above shows a mature sperm cell, magnified 5,000 times, burrowing into the tough outer membrane of an egg at the instant of conception. Chemical changes in the egg's membrane prevent any other sperm from entering. In a moment, this sperm—whose head contains the sperm cell's nucleus and its 23 chromosomes—will proceed to the egg's interior and fuse with its nucleus, creating an embryo with a full complement of 46 chromosomes. The sex of the child that will develop from that embryo depends on what type of sperm cell has fertilized the egg. A sperm carrying the Y chromosome will give rise to an XY combination of sex chromosomes *(bottom left)*, resulting in a baby boy. But a sperm with the X chromosome yields an embryo with two X chromosomes *(bottom right)* and a baby girl.

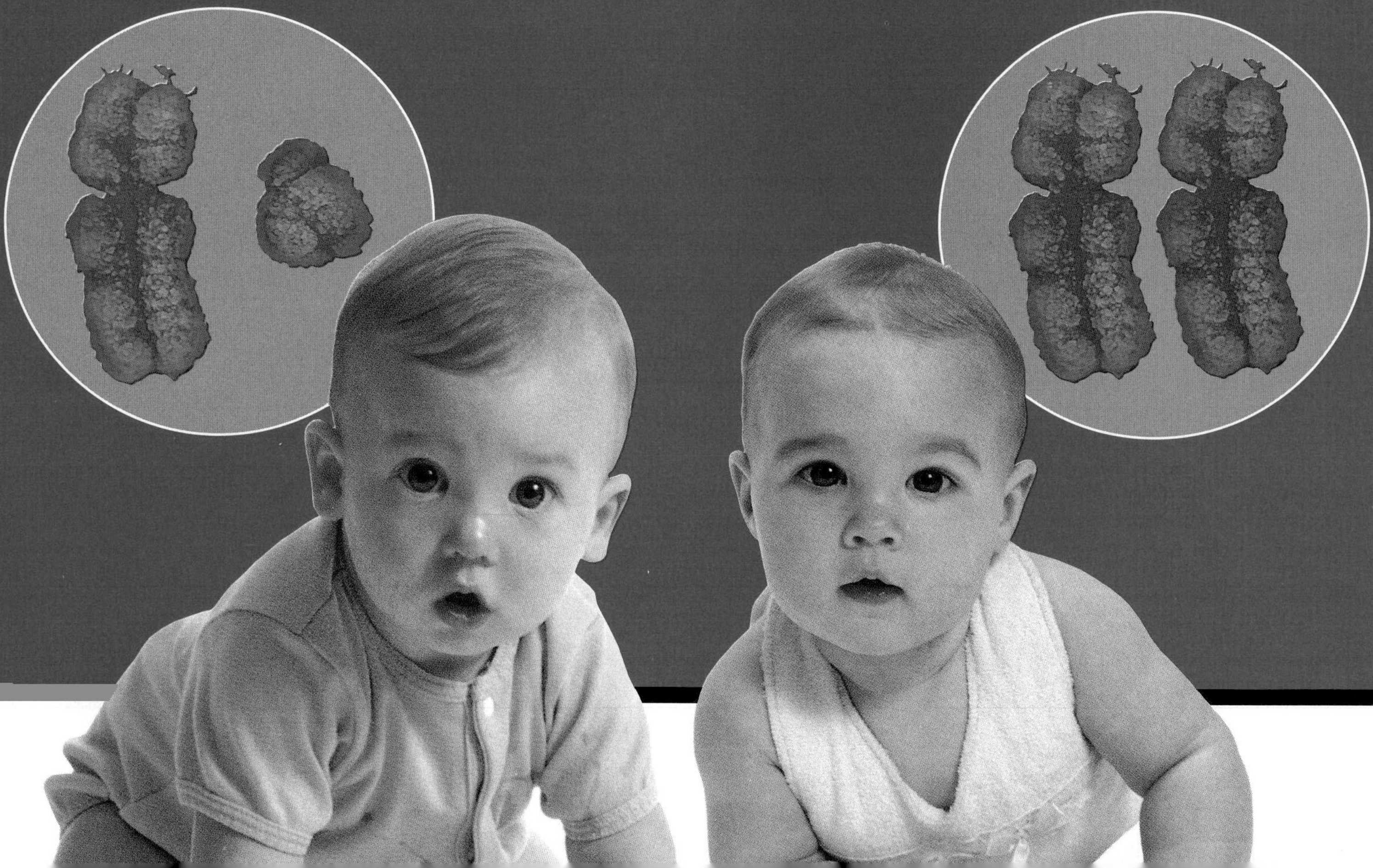

genes working together. The complexity arises from the fact that genes carry instructions for making proteins that sometimes spread throughout an organism and either have multiple effects on the organism's form and function or work in combination with other proteins—stemming from other genes—to produce a single effect.

Although some investigators took up the challenge of puzzling out such intricacies, many chose to focus on the more straightforward dominant and recessive patterns as they manifest themselves in humans, particularly in cases where things seem to have gone wrong and a disease becomes part of a family's genetic heritage.

There are, of course, a tremendous number of variations in the genetic program that are entirely benign, whether the trait in question is dominant or recessive. For example, "achoo" syndrome, or photic sneezing, is a common dominant condition that merely causes a person to sneeze when exposed suddenly to bright light. Other inconsequential dominant traits include curly hair, a second toe longer than the big toe, and hair growth between the second and third knuckles of the fingers. Among harmless recessive traits are blond hair, a second finger longer than the fourth, and an absence of dimples in the cheeks.

A small but measurable percentage of the time, however, mutations in one or both alleles of a particular gene alter the program in harmful ways, causing wasting ailments or even death. The dynamics in such cases are somewhat different from those characterizing normal dominant and recessive traits.

For one thing, malign traits that are dominant seldom occur in homozygous form—that is, with two copies of the responsible allele present—because such conditions tend to be quite rare in the first place, making it unlikely that two people affected by the condition will meet and have children with the potential to inherit two mutated alleles. But when the unlikely does indeed happen, the negative effects are so potent that the consequences are nearly always dire. For example, fetuses with a double dose of the genetic error that causes a certain type of dwarfism are almost never carried to term and, if they are, usually die in infancy.

Other dominant conditions prove lethal even in heterozygous individuals. One of the more tragic instances is Huntington's disease, which was first recognized in Europe, perhaps as early as 1600; it takes its name from the American physician who described its symptoms in 1872, after studying a Long Island family plagued by the disorder. Perhaps the most well known person to be stricken by Huntington's this century was the American folk singer Woody Guthrie, who began displaying symptoms at the height of his career and was several times misdiagnosed by doctors.

Misdiagnosis was not unusual because the disease can initially masquerade as a relatively common psychological problem such as depression, accompanied perhaps by mild facial twitches and a slight loss of coordination. But this soon leads to a decline in intellect and in bodily control, as well as violent muscle spasms that eventually make it impossible for the victim to swallow. Woody Guthrie experienced all this and worse, suffering storms of rage and periods of deep depression that finally forced his family to have him hospitalized. Even then his condition was not correctly identified for a number of years. He died in 1967 at age 55.

Of all its horrors, one of Huntington's most distressing characteristics is how long it takes to manifest itself. In many cases, including Woody Guthrie's, the affected individual may already have had children before

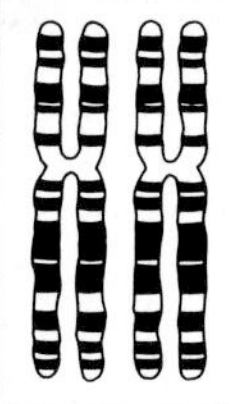

A Chromosomal Lineup

Many diseases are the result of chromosomal damage that may, for example, delete or rearrange portions of DNA; or a pair of chromosomes may fail to separate at the appropriate time during sex-cell formation, resulting in an excess or deletion of genetic material. Geneticists can often identify chromosomal abnormalities—during prenatal testing, for example—by looking at a specialized arrangement of a person's chromosomes, known as a karyotype.

In a normal human karyotype (from *karyo*, Greek for "cell nucleus"), the 46 chromosomes—23 pairs—are ordered by size and other characteristics. Twenty-two pairs, numbered 1 through 22, are called autosomes; the sex chromosomes are the 23d pair. In theory, karyotypes are unique to each individual—except, perhaps, for identical twins—but characteristics such as banding patterns (created by dyes used in the karyotyping process) and the position of the chromosome's centromere, or primary constriction point, vary only slightly among healthy individuals. Geneticists have thus been able to create diagrammatic representations, or idiograms, of human male and human female karyotypes for comparison. (An idiogram of a pair of chromosome number 2 is shown above.) An individual's karyotype is derived from a photograph of chromosomes isolated from a cell undergoing division. By counting the number of chromosomes, geneticists can spot such abnormalities as trisomy 21, or Down syndrome, which occurs when a person has three copies, instead of two, of chromosome 21. More subtle rearrangements of genetic material require comparison of the karyotype with an idiogram.

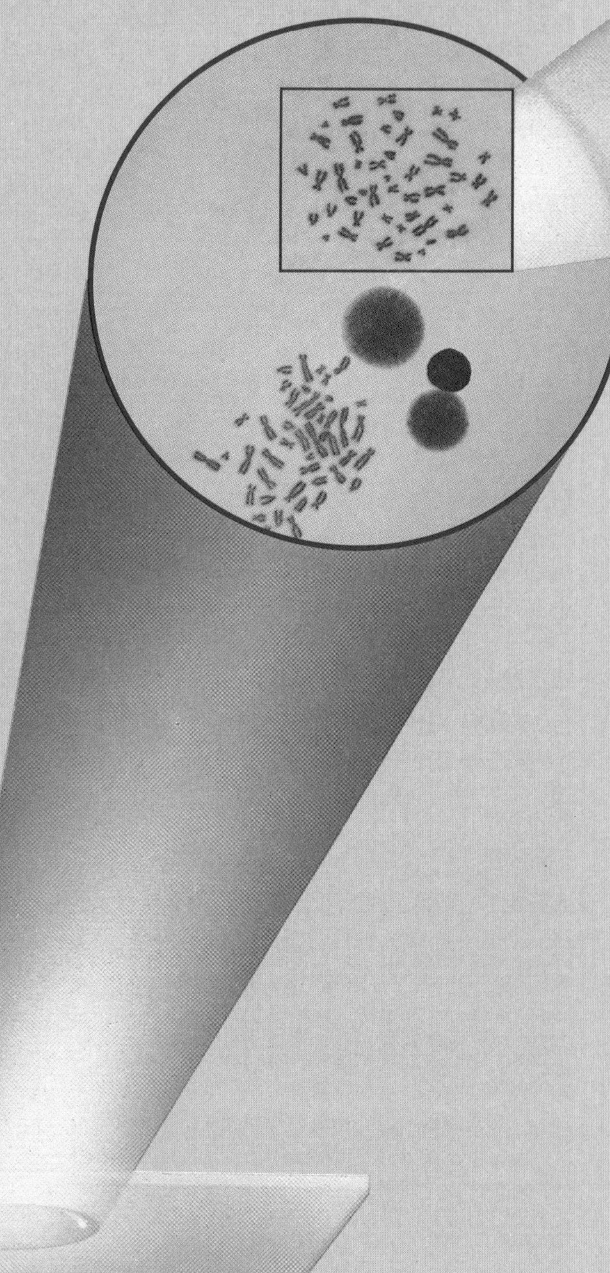

1 PREPARING THE SLIDE. Chromosomes used to create a karyotype are isolated from cells undergoing division, usually from blood or tissue samples, or from amniotic fluid. The addition of a chemical halts cell division at the stage in which chromosomes are most readily visible, and then the sample is placed on a microscope slide.

2 UNDER THE MICROSCOPE. Physical pressure or another chemical solution disrupts the cell nuclei and releases the chromosomes. Two sets of dispersed chromosomes and several intact cell nuclei are shown below. The chromosomes are then stained with dyes that produce characteristic banding patterns.

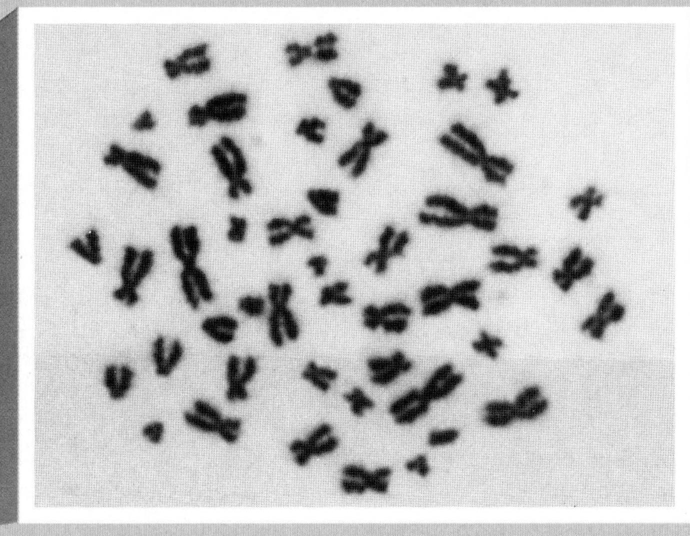

3 A CLOSEUP VIEW. A photograph is then made of the chromosome group with the most visible banding patterns *(left)*. When the photograph is enlarged, each chromosome's banding patterns, the length of its two segments, and the position of its centromere can be seen with the naked eye. The photograph is then cut apart and the chromosome pairs matched up in size order to create a karyotype *(below)*.

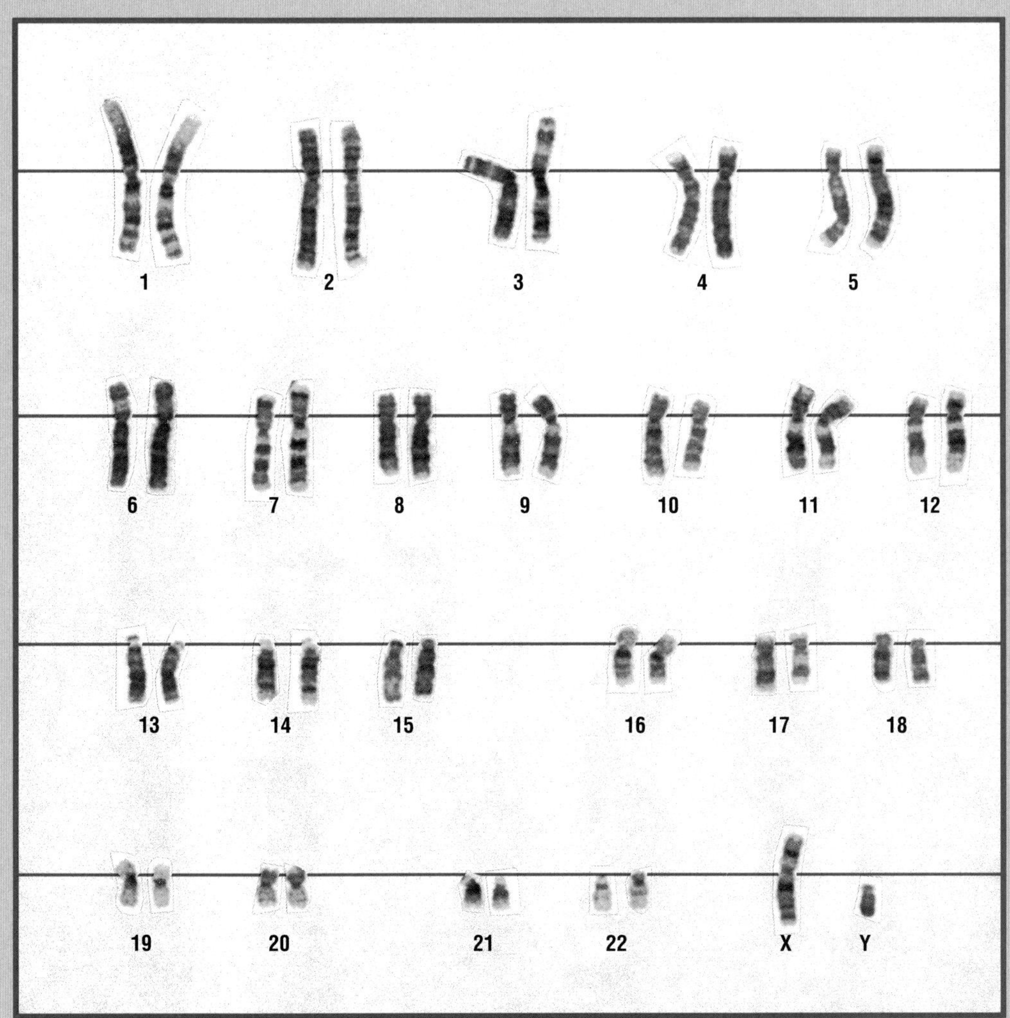

4 A SAMPLE KARYOTYPE. This karyotype of a normal male was created by arranging the 23 pairs of human chromosomes from largest to smallest. For chromosomes of nearly equal size, the banding pattern is used to differentiate between pairs, such as between pair 21 and pair 22. Although karyotyping can be automated, computers are rarely able to discern the subtle differences well enough to arrange all 46 chromosomes in the proper order.

developing any symptoms. And because the condition is dominant, each offspring has a 50 percent chance of receiving the fatal allele and eventually developing the disease.

A second major class of inherited disorders are those that follow recessive patterns. Because they only become expressed when identical copies of a mutated allele are delivered to an offspring, recessive traits may slip unnoticed through several generations, masked by an alternative dominant trait. Tracing the inheritance of recessive traits is thus something of a sleuthing job best performed on families with well-documented pedigrees. Genetics textbooks are full of case studies of European royalty, the Amish of Pennsylvania, and other isolated ethnic or social groups.

The numerics of chance are somewhat more intricate for recessive disorders than for dominant ones, but the rules established by Mendel are still a good guide. An individual who has only one copy of a harmful allele for a recessive condition is a carrier for that condition and has equal odds of passing the allele on to a child. A child who receives that allele will not develop the disease so long as the other parent supplies a benign version of the same gene. Even if both parents carry the harmful allele, the children are not necessarily doomed to develop the disorder: Only one in four, on average, will inherit two copies of the recessive-trait allele; another quarter will receive two harmless dominant alleles; and half will receive one of each, becoming carriers like their parents.

Estimates vary, but humans probably inherit somewhere between three and seven recessive alleles that would prove lethal if present in a double dose. Thus it is that healthy parents can produce children with fatal recessive conditions such as cystic fibrosis, which causes massive buildups of mucus especially in the lungs, giving rise to dangerous infections that eventually weaken the lungs to the point of failure.

Other recessive disorders tend to appear in close-knit ethnic populations, where the chances are greater that both parents will carry the same recessive allele and so give birth to an affected child. One such disease is Tay-Sachs, a horrific ailment that primarily affects Jews of central or eastern European descent. Working with lethal swiftness, it blinds and paralyzes its victims, who usually die before the age of four. As a result, the disease should become less prevalent over time: Because its victims cannot produce offspring, the number of copies of the mutated gene in a given population should continually decrease. Yet Tay-Sachs disease has somehow managed to remain a curse for generations.

Researchers have come up with several different explanations for the unusual survival of Tay-Sachs, including that fresh mutations of the appropriate gene take place relatively frequently, in effect reintroducing the disease time and again. But a more likely cause traces its persistence to a related benefit. It seems that the same mutated allele that kills when an individual possesses two copies may offer protection against tuberculosis when only one copy is present. In eastern European cities earlier this century, tuberculosis was responsible for as many as one-fifth of all deaths. But among the Jewish population, the death rate from tuberculosis was much lower. Although theorists cannot be absolutely sure, they strongly suspect that the Tay-Sachs allele conveyed some resistance to tuberculosis, protecting those who carried it and so, by natural selection, increasing its proportion in the population. A similar situation has been discovered with the blood disease sickle cell anemia, which affects people of African descent and populations in Arabia, India, and around the Mediterranean. It is usually deadly when two

AN ENDURING ABNORMALITY. Geneticists believe this statuette from the second century BC depicts a woman suffering from a kind of dwarfism still common today. Because the trait is dominant, only one copy of the responsible gene need be present for a person to be affected. In turn, that person's offspring will have a 50 percent chance of inheriting the disorder.

copies of its allele are present; in carriers, however, it has been shown to provide immunity to malaria, a ravaging affliction occurring in precisely those regions where sickle cell anemia is most common.

Determining whether a given trait is dominant or recessive requires no knowledge of which chromosomes hold the genes themselves. But by their very nature, the patterns of inheritance for one particular class of traits reveal where the responsible genes reside: on the sex chromosomes. Examples of such traits were originally identified in humans not long after Thomas Morgan's discovery of X-linked traits among fruit flies. Their numbers are relatively few compared to those associated with all the other chromosome pairs—a few hundred as opposed to several thousand—and almost all sex-linked traits trace to the X chromosome. In fact, only two traits have been linked to the Y chromosome, one of which is maleness itself and the other a tendency to grow hair on the outer rim of the ear.

Human X-linked traits, like those in other species, come in both dominant and recessive forms, their specific patterns painstakingly worked out primarily through careful analysis of family trees. A given trait can be identified as dominant, for example, through the progeny of an affected male. All his daughters will be affected because they invariably receive his only X chromosome, where the responsible allele resides; and unless his wife is also affected, all his sons will be free of the trait because they receive only his Y chromosome. The pattern is different for an affected female, who typically will have one mutant and one normal allele. Since she can pass either of her X chromosomes to both male and female offspring, all of her children—regardless of their sex—stand a 50 percent chance of exhibiting the trait; if her husband is also affected, of course, the chance for daughters to be affected rises dramatically.

For recessive X-linked disorders to show up in a woman requires the presence of two defective alleles, but just one will cause the disorder in a man, who has no second X chromosome to

Mixing and Matching: Patterns of Inheritance

The combination of features that makes each of us an individual—almond-shaped eyes, say, and a bump on the bridge of the nose, or any one of almost 6,000 characteristics so far identified by geneticists—is the product of the mixing and matching of genetic instructions passed along from our parents. The instructions for a given characteristic are carried in the form of alleles, or copies of the same gene. As illustrated at right, each parent has two alleles and passes one along to each child. In most cases, the allele that would result in a dominant trait will override the allele responsible for a recessive one. The recessive trait is not eradicated, however, but can be handed down; its effects will reemerge when it is paired with a recessive allele in a future generation.

These rules of dominance do not always hold, however, especially in the case of recessive X-linked traits, those associated with one of the sex chromosomes. Females who carry an abnormal allele on one of their two X chromosomes often do not display the trait because it is masked by the dominant normal allele on their other X chromosome. But a carrier mother can transmit the abnormal allele to a son, who inherits his only X chromosome from her; with no countervailing allele on his Y chromosome, the son will display the usually recessive X-linked trait.

In the drawings at right, representing the statistical distribution of dominant, recessive, and X-linked recessive traits, the alleles are denoted as letters of the alphabet—uppercase for dominant, lowercase for recessive. Letters with an orange background indicate the alleles necessary for a trait's physical expression.

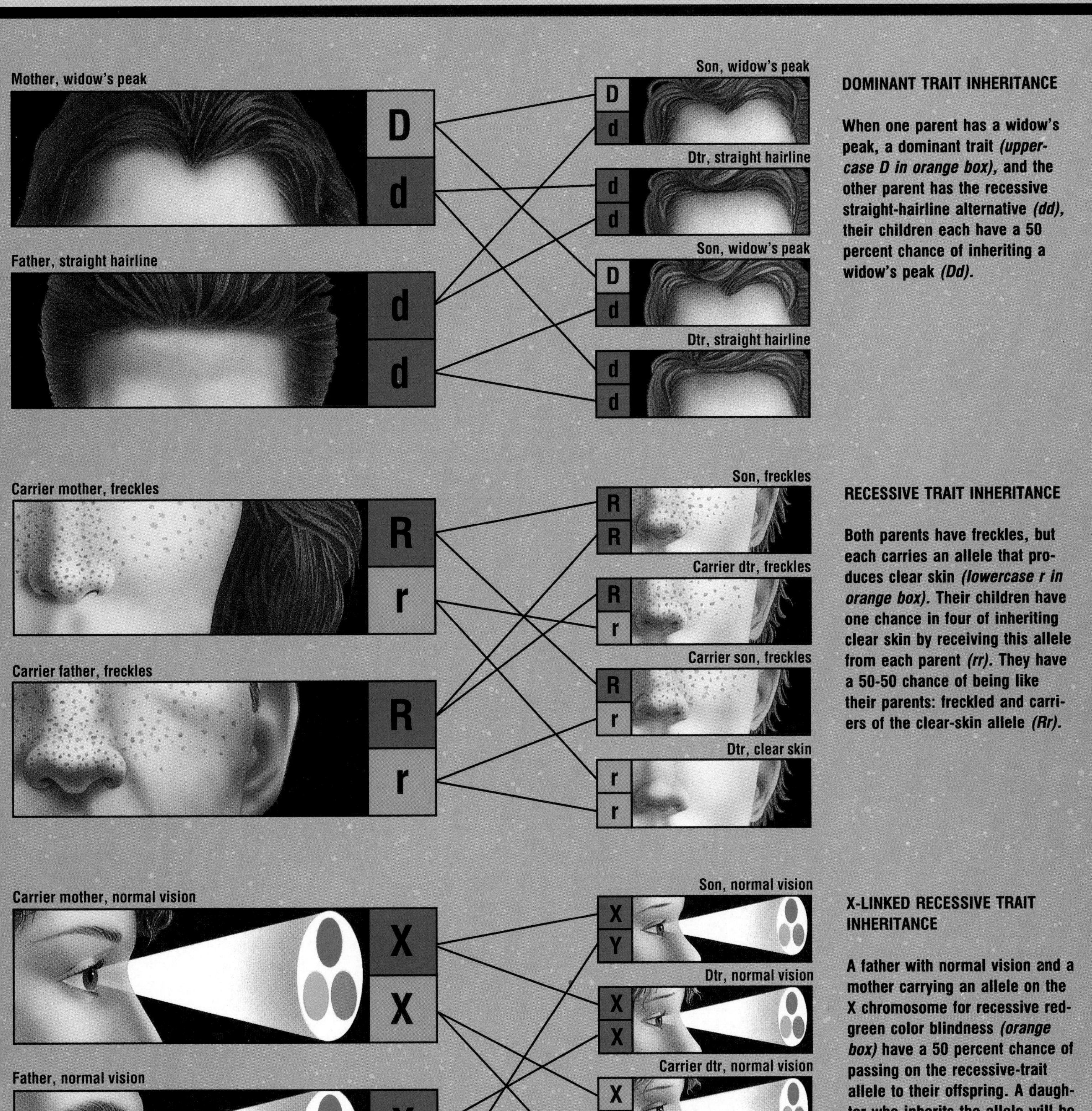

DOMINANT TRAIT INHERITANCE

When one parent has a widow's peak, a dominant trait *(uppercase D in orange box),* and the other parent has the recessive straight-hairline alternative *(dd),* their children each have a 50 percent chance of inheriting a widow's peak *(Dd).*

RECESSIVE TRAIT INHERITANCE

Both parents have freckles, but each carries an allele that produces clear skin *(lowercase r in orange box).* Their children have one chance in four of inheriting clear skin by receiving this allele from each parent *(rr).* They have a 50-50 chance of being like their parents: freckled and carriers of the clear-skin allele *(Rr).*

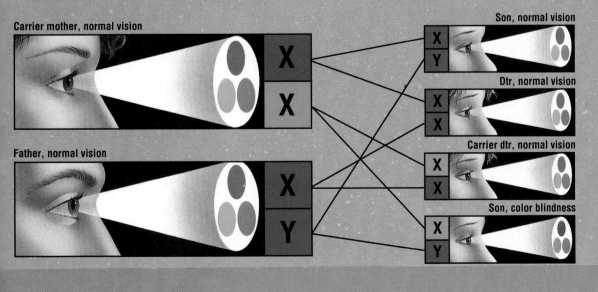

X-LINKED RECESSIVE TRAIT INHERITANCE

A father with normal vision and a mother carrying an allele on the X chromosome for recessive red-green color blindness *(orange box)* have a 50 percent chance of passing on the recessive-trait allele to their offspring. A daughter who inherits the allele will be a carrier; a son will express the trait and see shades of gray and black instead of red and green.

carry a possible counterbalancing normal allele. Therefore, just as with Morgan's white-eyed fruit flies, most affected individuals turn out to be males, and all the daughters of an affected father become carriers because they receive one of their two X chromosomes from him.

The American biologist E. B. Wilson, a colleague of Morgan's at Columbia, was the first to recognize this unique pattern of inheritance in humans, just shortly after Morgan's own discovery with the fruit flies. Wilson did so by studying several generations of families afflicted with red-green color blindness (*page* 33). Researchers soon realized that more threatening diseases were also in this category, most notably hemophilia (*pages* 36-37), which had become the object of much study in the late 19th and early 20th centuries, because it repeatedly cropped up in the bloodlines of European royalty.

Sex-linked traits more than any others proved that a great deal could be deduced not only about the mechanisms of heredity but also about where certain genes are among the chromosomes purely from examinations of the patterns that traits exhibit across generations. This type of re-

SEXUAL AMBIGUITY. Spanish hurdler María Patiño never realized she was genetically male until she failed an athletic competition's gender test, which revealed that her cells contain the XY combination of sex chromosomes. She is, however, physically female—a result of her body's inability to respond to the testosterone it produces, causing the development of feminine features.

search confirmed once more the validity of Mendel's seminal experiments in the monastery garden. But the urge to fathom the secrets of inheritance at its deepest levels continued to grow stronger. Again and again through the first half of the 20th century, attention turned to the still-mysterious question of what genes are and how they work. The answer that finally emerged was stunning in its simplicity and in its elegance, and few would dispute its rank as the most important discovery in the entire history of biological science.

On the surface, identifying the makeup of genes might seem a straightforward matter. Since the mid-1800s, scientists had been developing techniques for analyzing the chemical composition of organic material and in some cases were able to construct models that illustrated how the various components of organic molecules fit together. Once it was determined that genes reside on the chromosomes, researchers applied these techniques to chromosome strands extracted from cells, hoping to pin down the chemical structure of the genes themselves.

But their results were equivocal. Chromosomes were found to consist of two types of biochemical material, protein and deoxyribonucleic acid, or DNA, both of which were present in about equal proportions. DNA was first discovered in 1869, and by 1940 chemists had managed to figure out its constituent parts: One unit of DNA, known as a nucleotide, included a phosphate bound to a sugar, which in turn formed a connection with a third component called a base, of which there were four types. Proteins were even more complex, consisting of 20 different subunits called amino acids. Although researchers had all the parts of these intricate molecules identified, they were unable to discern their three-dimensional structure—information that was essential to understanding their function. Nor could scientists tell which of the two was the raw material of genes.

Some investigators had their suspicions, however. Given how complex the genetic message had to be to carry all of an organism's biological instructions, DNA—which had only four variables—appeared to be an unlikely candidate. In fact, Max Delbrück, a molecular biologist very much involved in the research at the time, noted years later that most of his colleagues considered DNA to be "a stupid substance," incapable of directing the production of anything interesting. The accepted wisdom for a number of years, then, was that proteins must be the stuff of genes.

Yet others were not so ready to dismiss DNA just on theoretical grounds. In the early 1940s, a researcher named Oswald Avery, working at the Rockefeller Institute in New York, made a fascinating discovery in a series of experiments with bacteria. He was working with two strains, one of which caused pneumonia in laboratory mice, while the other was essentially harmless. He found that an extract produced by heating and killing off the deadly bacteria had the power to transform the harmless bacteria into killers. Furthermore, the offspring of the transformed bacteria continued to prove lethal, generation after generation. The obvious conclusion was that the extract contained something, which Avery initially called transforming principle, that had transferred genetic instructions from the first strain to the second. The transforming principle had to contain genes.

When Avery and his colleagues analyzed the extract, they found that it contained several compounds, including the two components of chromosomes—protein and DNA. This confirmed the supposition that there was genetic material present, but for the moment left unanswered the question

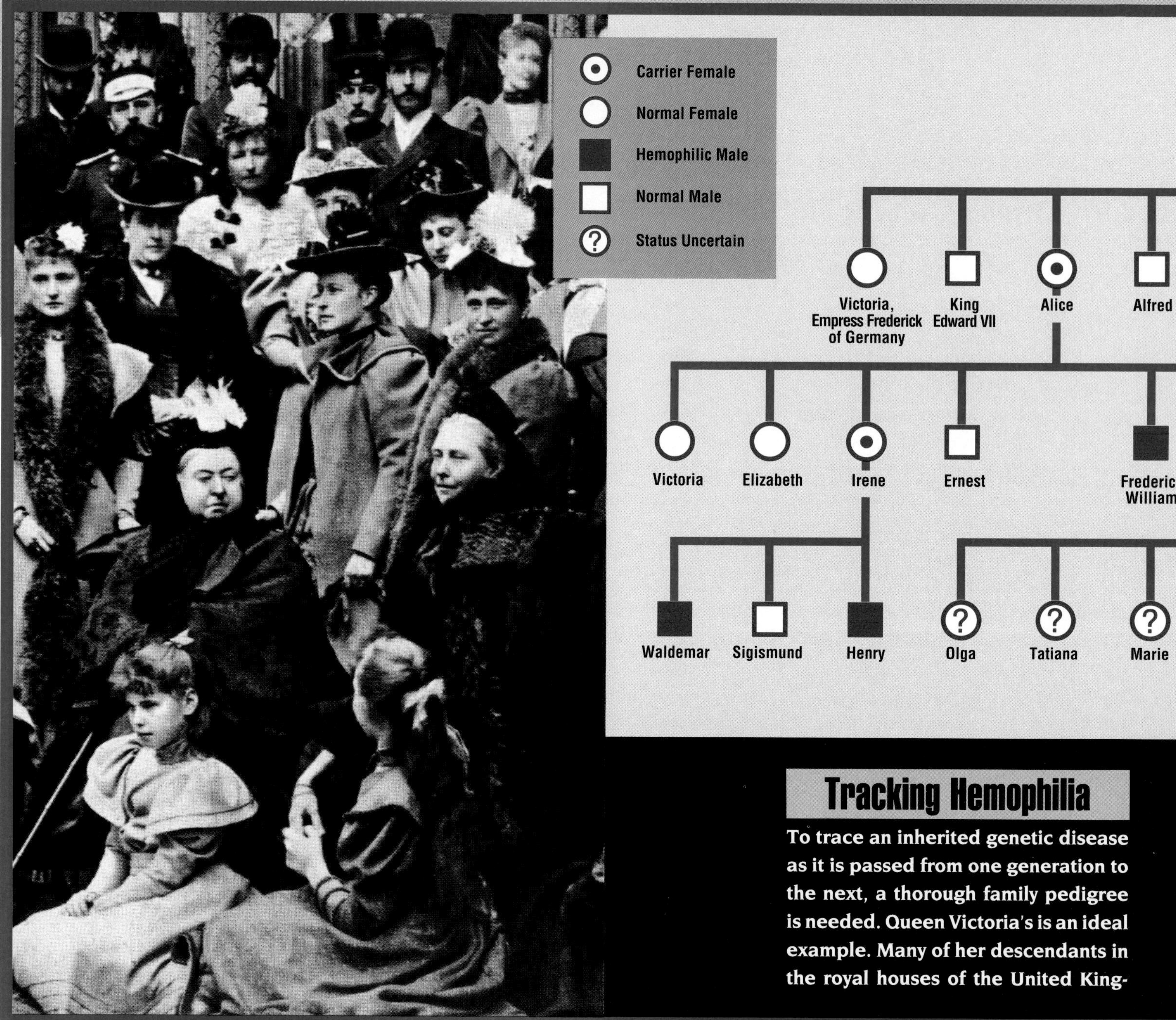

Tracking Hemophilia

To trace an inherited genetic disease as it is passed from one generation to the next, a thorough family pedigree is needed. Queen Victoria's is an ideal example. Many of her descendants in the royal houses of the United King-

of its makeup. So Avery's team went about purifying batches of the extract to isolate each of its constituent elements. When the protein was isolated and then injected into harmless bacteria, they remained harmless. But when the extract was purified so that only DNA remained, the extract's power to transform the harmless bacteria reappeared. The transforming principle had been identified: It was DNA that carried the genetic message.

Avery published his results in 1944, and although some researchers maintained that only proteins had the necessary complexity to encode genetic instructions, as more experiments confirmed his findings, the scientific community was soon convinced that genes are made of DNA.

The most crucial piece of evidence, however, was still to be uncovered. DNA's three-dimensional structure remained a mystery, and with it the secret of how DNA was able to replicate itself—the key mechanism of inheritance. By the early 1950s, two insights by scientists working independently offered tantalizing hints to the ultimate solution, which was just around the corner.

The first of these insights came

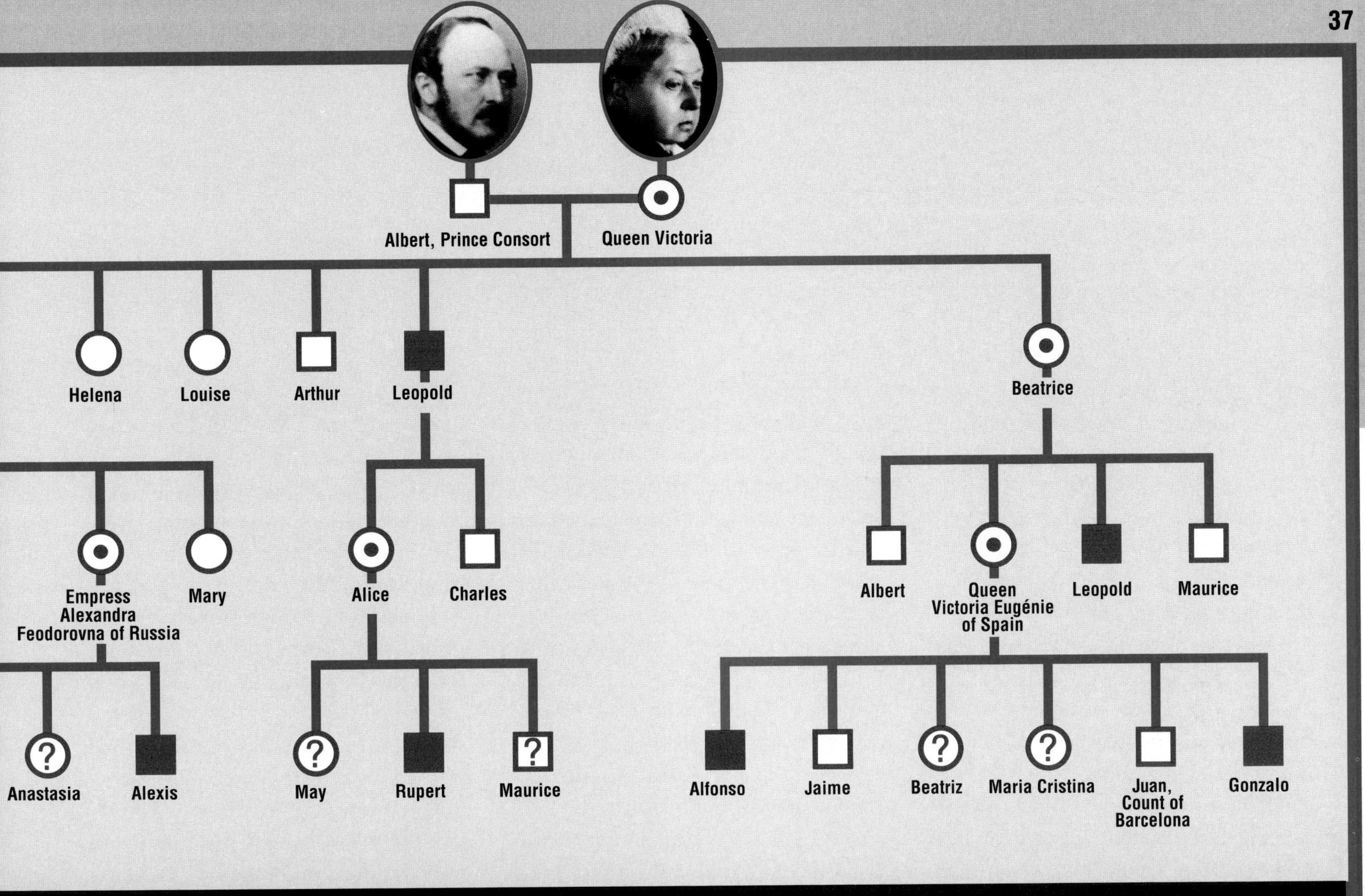

dom, Spain, and Russia were hemophiliacs or carriers of the allele that causes this blood-clotting disorder. Classic hemophilia is an X-linked recessive trait, so it affects men almost exclusively; in most cases a woman will express the trait only if both her X chromosomes carry the allele.

Victoria's son Leopold was the first hemophiliac to appear in her family. Since neither her husband nor any of her ancestors had hemophilia, geneticists surmise that a mutation in one of Victoria's X chromosomes made her a carrier. Her two carrier daughters each had a hemophilic son, and the trait was also passed on to four carrier granddaughters. Six great-grandsons were hemophilic; one great-grandson and four great-granddaughters died before their status could be determined. The status of the other great-granddaughters will be known if any of their descendants has hemophilia.

from Erwin Chargaff, a biochemist at Columbia University who had been inspired by Avery's work to focus his attention on DNA. As he later noted, "Avery gave us the first text of a new language, or rather he showed us where to look for it. I resolved to search for this text." His efforts led him to examine cellular material from various organisms and determine the proportions of DNA's four bases in each, which had been assumed to be equal. They were not; however, he found that no matter which organism he examined, adenine (A) and thymine (T) always came in the same amount, and so did guanine (G) and cytosine (C). Not quite sure what to make of this, he published the data anyway, establishing what became known as Chargaff's rules: A equals T, and G equals C.

Meanwhile, a Scottish chemist by the name of Alexander Todd had been trying to figure out how the units of DNA called nucleotides joined to form DNA molecules. Eventually in 1952, he worked out the details: The sugar-and-phosphate portions of nucleotides linked up to form a continuous strand, and the bases dangled from it, forming an apparently random sequence. It was a major discovery, earning him the Nobel Prize in chemistry five years later. But there was still no cry of "Eureka!" The arrangement he posited allowed for too many possibilities in DNA's three-dimensional structure. It was still not clear what shape the molecule actually took, and how that shape enabled it to duplicate itself.

Just the year before Todd's analysis, however, two scientists at Cambridge University in England had begun their own attempt to ferret out the secret of DNA's structure. Brought together by another colleague at Cambridge's Cavendish Physics Laboratory, the American biologist James Watson and the English physicist Francis Crick made the most of the clues that others had provided. In addition to Chargaff's work and later Todd's, they also made use of data collected by Maurice Wilkins and Rosalind Franklin, two researchers at a laboratory in London; Wilkins and Franklin had been employing a technique for studying the structure of molecules—including DNA—by bombarding them with x-rays and analyzing the patterns of the deflected rays.

Using labeled bits of cardboard to construct a model, Watson and Crick tinkered with various arrangements of DNA's known parts, basing their attempts primarily on Franklin's x-ray images. Then, on the morning of February 28, 1953, the final piece of the puzzle clicked into place. Watson realized that Chargaff's rules—A equals T, G equals C—made sense if the members of each pair formed chemical bonds with each other. A DNA molecule was not a single one of Todd's strands, but two strands, linked by their complementary bases. The x-ray data told Watson and Crick that the strands most likely formed a double helix—a spiraling shape like a twisted ladder. Together they quickly tried out the pattern with the cardboard model, and it worked. They rushed off to celebrate at a local pub, where Crick is reported to have announced to the crowd, "We have discovered the secret of life!"

The key feature of their model—what made it have such an immediate impact—was that the form of the molecule revealed how DNA performed its most vital function, replicating itself. Because an A would bond only with a T, and a G only with a C, the sequence of bases along one strand was always an exact complement of the sequence on the other. Unzipping the two strands would thus yield two halves, each bearing all the information necessary for the creation of a new complementary strand, to form a duplicate of the original molecule.

In addition, though there were only four types of bases, they could be arranged in a virtually limitless variety of sequences along DNA's chainlike strands, meaning DNA obviously had the ability to carry complex genetic information. Crick perhaps stated it best in a letter to his son less than a month after the discovery. "Our structure is very beautiful," he wrote; "it is like a code. If you are given one set of letters you can write down the others. Now we believe that the DNA *is* a code. That is, the order of bases (the letters) makes one gene different from another gene."

In the wake of Watson and Crick's great breakthrough, further insights emerged apace. Crick himself contributed to the discovery of how the DNA code is actually put to work to produce the specific substances that form an organism and regulate its biological activity (*pages* 39-49). But most important of all, a profound shift had taken place in the examination of the human blueprint. Investigators now had an exciting new way of pursuing the secrets of inheritance, by looking at the actual details of the genetic text—the letters, words, and sentences in which heredity is written.

CARRYING OUR GENETIC LEGACY

All cells have the chromosomes that carry a cell's genetic heritage to its children as DNA. But only for a short period during cell division do chromosomes take on the classic X-shape. The rest of the time, they are strung out much more loosely inside the cell nucleus. In this form, called chromatin, DNA and the specific sequences that constitute genes serve the cell in two vital day-to-day activities.

The less-frequent task is cell division, for which a perfect copy of genes must be created. But far more often, genes are called on to produce proteins. Indeed, from one point of view, a cell is little more than a busy protein factory. Virtually everything in the body contains one or more kinds of protein molecules, each of which is modeled on an individual gene. Proteins can be found in the membranes of bone and other cells, in the juices of the digestive system, and as disease-fighting antibodies like the molecule shown here. How DNA replicates itself and how a cell manufactures proteins—both explained on the following pages—are marvels of life's microscopic ingenuity.

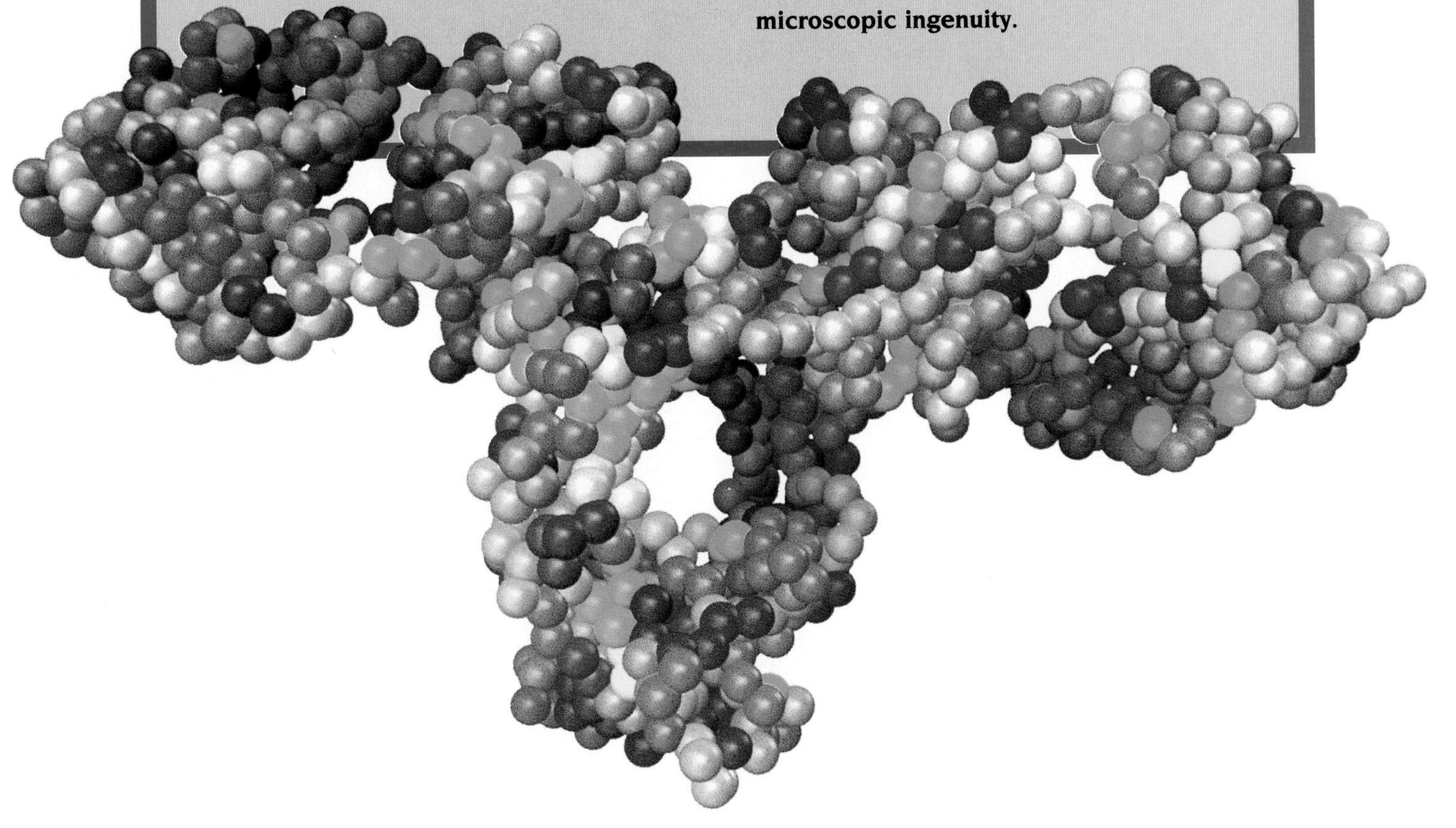

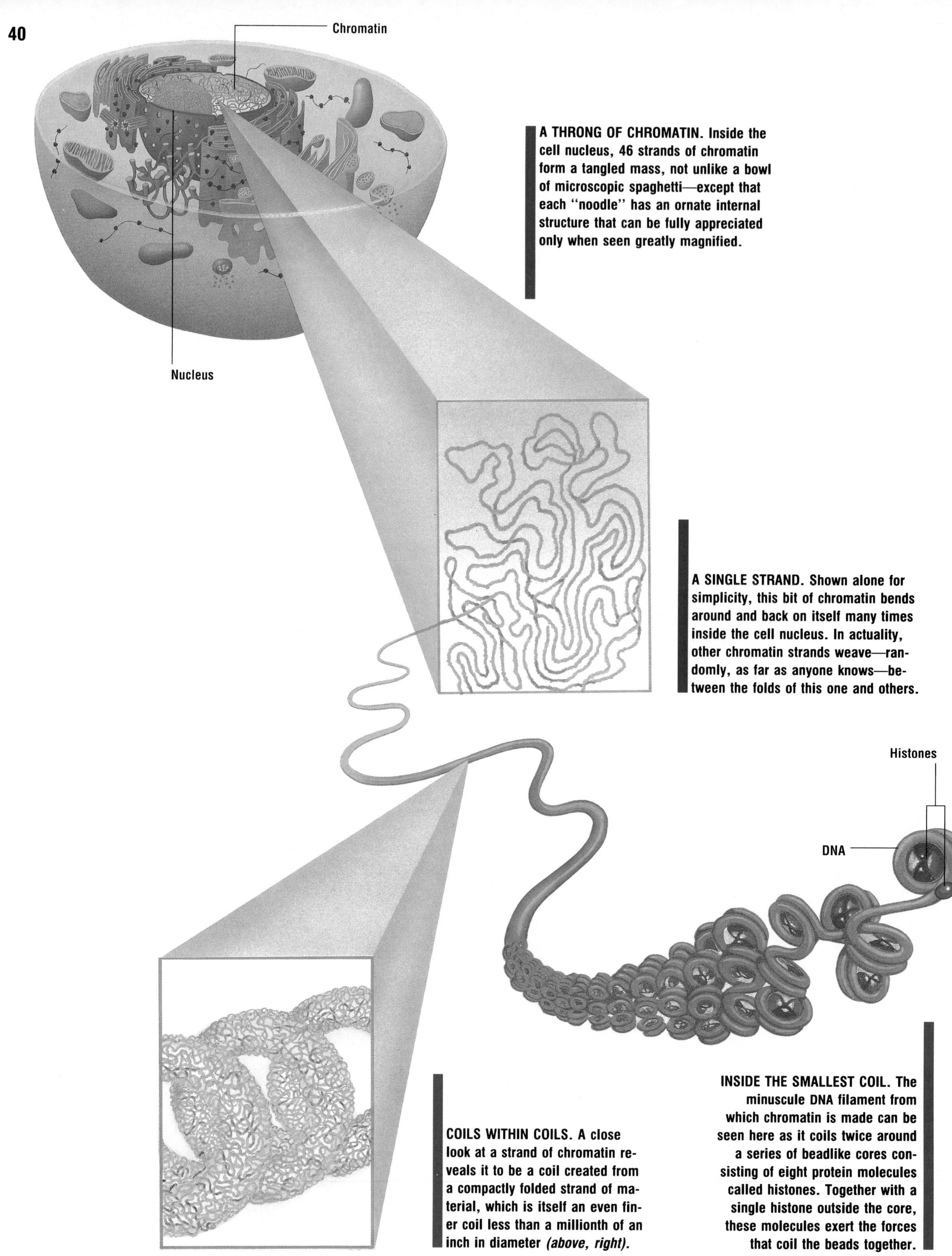

A THRONG OF CHROMATIN. Inside the cell nucleus, 46 strands of chromatin form a tangled mass, not unlike a bowl of microscopic spaghetti—except that each "noodle" has an ornate internal structure that can be fully appreciated only when seen greatly magnified.

A SINGLE STRAND. Shown alone for simplicity, this bit of chromatin bends around and back on itself many times inside the cell nucleus. In actuality, other chromatin strands weave—randomly, as far as anyone knows—between the folds of this one and others.

COILS WITHIN COILS. A close look at a strand of chromatin reveals it to be a coil created from a compactly folded strand of material, which is itself an even finer coil less than a millionth of an inch in diameter *(above, right)*.

INSIDE THE SMALLEST COIL. The minuscule DNA filament from which chromatin is made can be seen here as it coils twice around a series of beadlike cores consisting of eight protein molecules called histones. Together with a single histone outside the core, these molecules exert the forces that coil the beads together.

COILED AND COMPACT FOR EFFICIENCY

In order for the tens of thousands of genes in the human genome to fit comfortably inside a cell nucleus a mere six-thousandths of a millimeter in diameter, they must be packed together very efficiently.

The stratagem for doing so begins with a simple twisting of the DNA strand, shortening it modestly. Then come multiple coilings that further shorten the package and thicken it into a comparatively chunky filament of chromatin. Some 15 times shorter than the DNA strand in its relatively fragile, uncoiled state and 250 times thicker, chromatin is less susceptible to damage inside the nucleus.

Often during the life cycle of a cell, each chromatin strand must, in essence, let down its defenses and uncoil to perform its two crucial duties. For example, strands relax to begin the process of making the proteins that are required by the functions of particular cells (*pages* 44-45). They also uncoil in order to make copies of themselves as a cell heads toward division, at which point chromatin takes on the even more compact, X-shaped form of the chromosome.

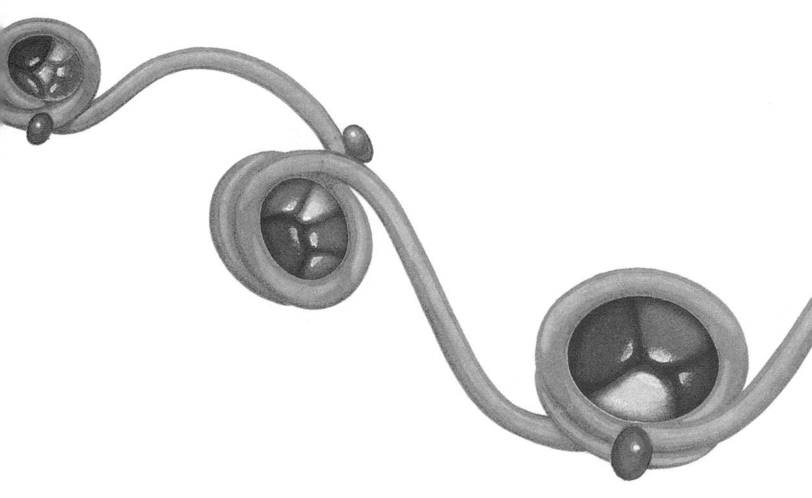

THE ORIGINAL TWIST. A closer look at the strand of DNA *(above, right)* clarifies its structure—a ladder shape consisting of two twisted side rails linked by innumerable chemical rungs.

DNA BUILDING BLOCKS. Nucleotides, shown here as T-shaped elements, consist of a sugar-and-phosphate top attached to a base, in this case adenine *(purple).* Adenine nucleotides can link up only with others having a thymine base *(lavender),* and a guanine nucleotide *(dark green)* can have only a cytosine mate *(light green).*

REPLICATION BEGINS. Before cell division, a DNA molecule is unzipped by an enzyme known as DNA helicase, which severs the bonds between nucleotide base pairs. Two parent strands result, each of which serves as a template for the creation of a complete daughter strand of DNA.

ASSEMBLING DAUGHTER STRANDS. The new strands are created with the help of the enzyme DNA polymerase, which gathers loose nucleotides and attaches them to the two parent strands. On one strand, the DNA polymerase follows closely behind the unzipping enzyme as it progresses along the DNA molecule *(above).* However, for complex reasons having to do with a directionality inherent in DNA, the enzyme must labor in the opposite direction along the other strand *(below).* To do so, many polymerase molecules take turns at the work. Each assembles a small segment of the daughter strand, beginning near the point of unzipping and working toward a segment already completed by an earlier DNA polymerase molecule.

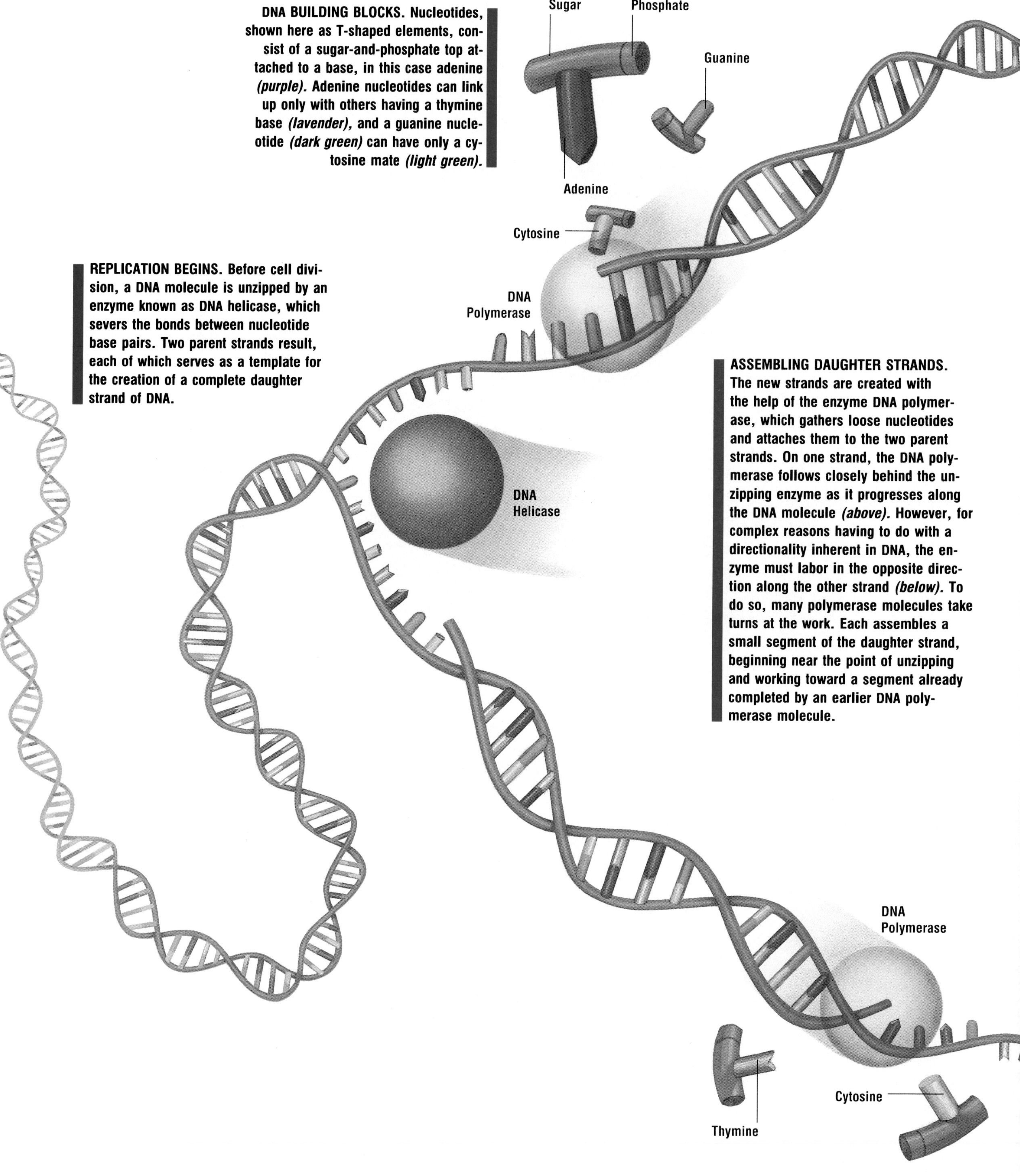

TWO DAUGHTERS FROM COMPLEMENTARY PARENTS

For a molecule so rich in information, DNA has a surprisingly simple structure that, barring accidents, permits flawless replication. This structure is based on four molecules called nucleotides, produced in large quantities by enzymes outside the nucleus, in the cell's cytoplasm (*pages* 18-19). These molecules join together in so-called base pairs to form the rungs and side rails of the DNA ladder.

As shown in the DNA strand illustrated here, only four arrangements of the nucleotide quartet are possible: two unique pairs, each base of which can fall on either side of the DNA strand. These four elements, arranged in long chains, offer all the variety needed to construct the many thousands of one-of-a-kind genes.

In essence, the DNA replication that precedes cell division involves splitting the strand of DNA down the middle into a pair of half strands called parents, then using each as a template for a complete DNA molecule known as a daughter. Because parent strands tend to reunite soon after unzipping, replication must proceed quickly—at the rate of about 2,000 rungs per minute. Errors are rare because both parents contain the original sequence of nucleotides, each of which will accept only the one that it had been linked with in the original strand of DNA.

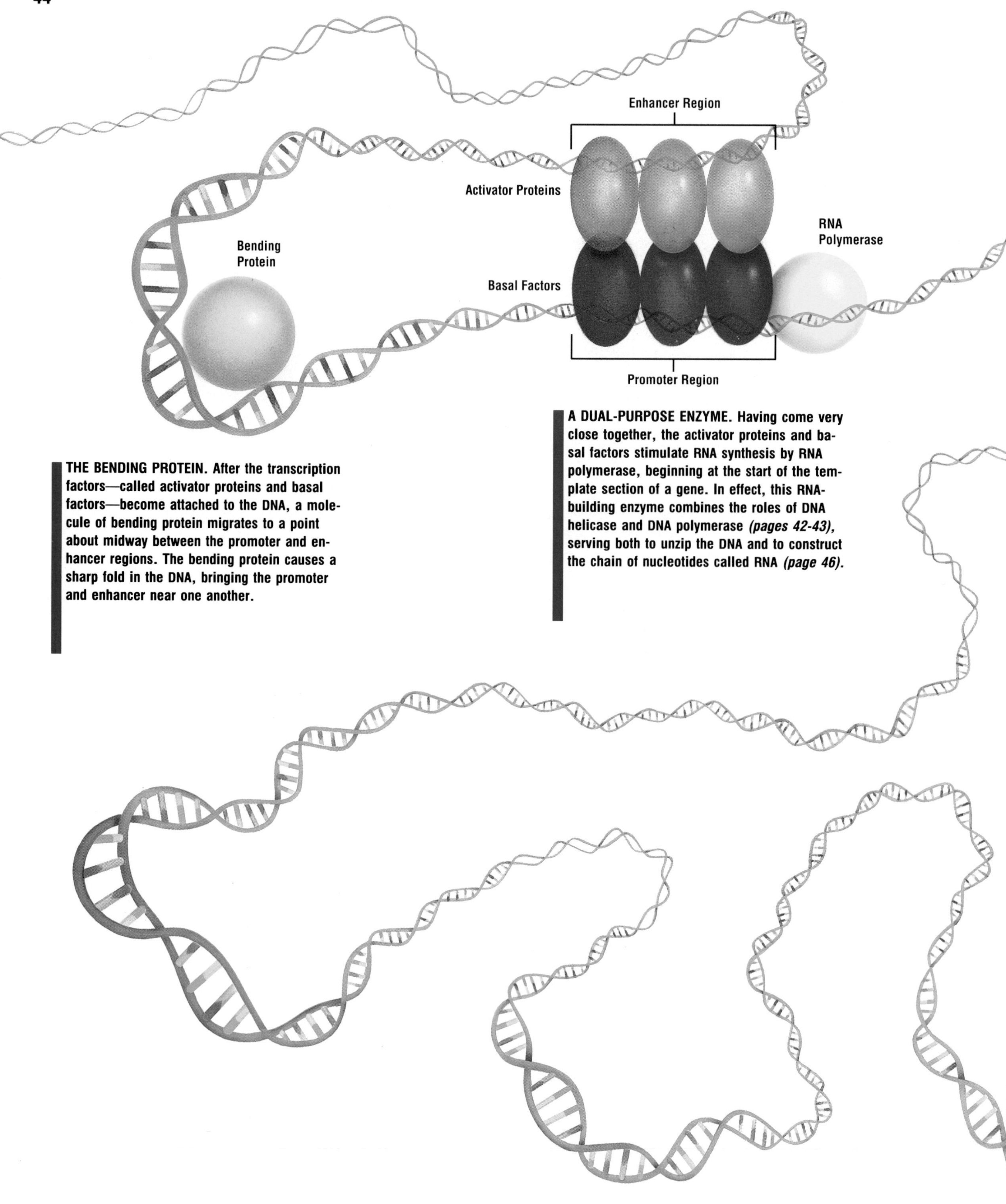

THE BENDING PROTEIN. After the transcription factors—called activator proteins and basal factors—become attached to the DNA, a molecule of bending protein migrates to a point about midway between the promoter and enhancer regions. The bending protein causes a sharp fold in the DNA, bringing the promoter and enhancer near one another.

A DUAL-PURPOSE ENZYME. Having come very close together, the activator proteins and basal factors stimulate RNA synthesis by RNA polymerase, beginning at the start of the template section of a gene. In effect, this RNA-building enzyme combines the roles of DNA helicase and DNA polymerase *(pages 42-43)*, serving both to unzip the DNA and to construct the chain of nucleotides called RNA *(page 46)*.

DNA'S ROLE IN BUILDING PROTEINS

A typical protein has a useful life of just two days. Consequently, new protein molecules must be synthesized constantly to replenish the supply of these vital compounds.

DNA's role in the process is to serve as a template for the assembly of nucleotides into ribonucleic acid, or RNA. In its final form, RNA itself becomes a template for assembling a chain of the amino acids that, in various combinations and sequences, make up all proteins.

This protein-building process is called gene expression by geneticists. Although the DNA of all cells has genes for all the body's proteins, no cell expresses every gene. Those in the retina, for example, are the only ones that produce proteins essential to the eye's light sensitivity, although the genes for those proteins are found in the DNA of liver cells, muscle cells, and other cells throughout the body.

Gene expression starts at two segments of a DNA strand called the promoter and enhancer regions that vary between 20 and 100 base pairs in length. The promoter region marks the beginning of a gene's RNA template section, shown snaking off the page at right. The enhancer region may lie thousands of base pairs away.

When a cell is called upon to produce one of its proteins, substances called transcription factors attach to the promoter and enhancer regions, kicking off a chain of events that begins with the bending of the DNA strand, shown at left.

MAKING RNA. The enzyme RNA polymerase unzips the DNA double helix (it rezips spontaneously after the enzyme passes) and links free-floating nucleotides into a strand of RNA. An encounter with a cytosine nucleotide in the DNA adds a guanine to the RNA chain and vice versa. A thymine nucleotide lengthens the chain with an adenine, but when RNA polymerase finds an adenine, it hooks a uracil nucleotide *(pink)* onto the chain.

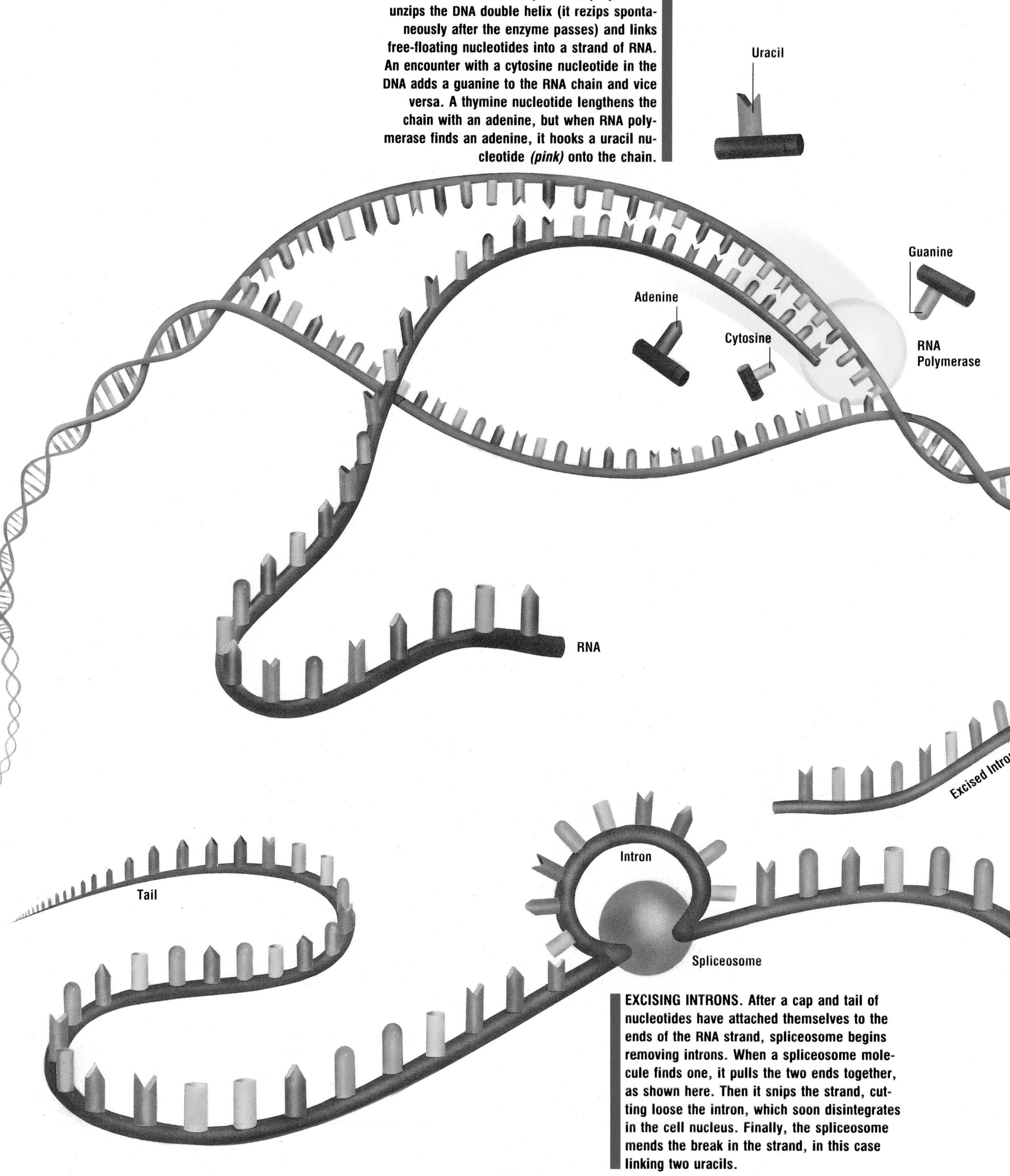

EXCISING INTRONS. After a cap and tail of nucleotides have attached themselves to the ends of the RNA strand, spliceosome begins removing introns. When a spliceosome molecule finds one, it pulls the two ends together, as shown here. Then it snips the strand, cutting loose the intron, which soon disintegrates in the cell nucleus. Finally, the spliceosome mends the break in the strand, in this case linking two uracils.

GENERATING AND REFINING THE GENETIC MESSAGE

Once activated by the reaction between the promoter and enhancer regions, RNA polymerase begins to unzip the gene's template section, using it to assemble a strand of RNA from free-floating nucleotides (*left*). The enzyme knows which side of the double helix to use because of DNA's inherent directionality.

So that RNA cannot be mistaken for DNA inside the nucleus, RNA nucleotides differ significantly from those found in DNA. For example, the sugar attached to the phosphate in each RNA nucleotide is called ribose. It has different numbers of hydrogen and oxygen atoms from the deoxyribose found in DNA nucleotides. In addition, a uracil base takes the place of DNA's thymine base.

For reasons not fully understood, some 70 percent of the nucleotides in a strand of RNA assembled to DNA specifications are superfluous. Getting rid of these meaningless nucleotides, known as introns, is the responsibility of a substance called spliceosome (*below, left*).

When this editing process is complete, the RNA strand gets a new name: messenger RNA, or mRNA. It is made up of hundreds of three-nucleotide units called codons—plus additional nucleotides in a cap and tail. In this form, the mRNA passes cap first through a special pore in the wall of the nucleus (*below, right*) and into the cytoplasm, where it becomes the template for assembling a protein (*overleaf*).

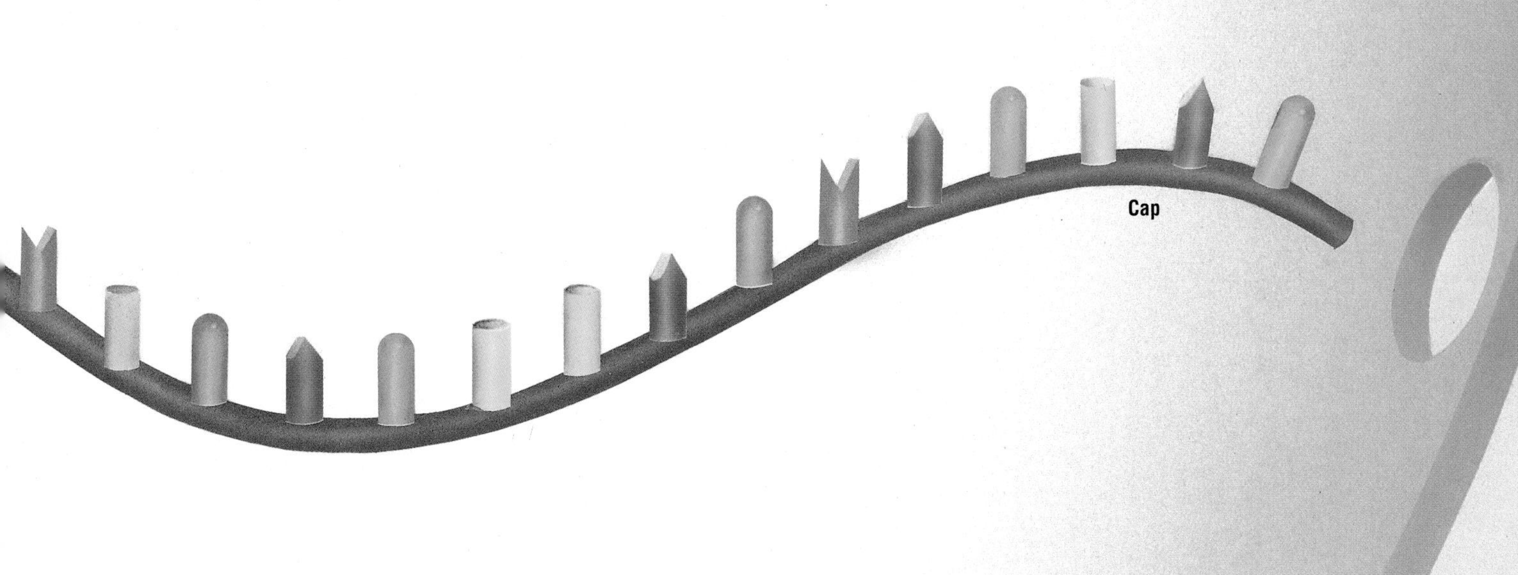

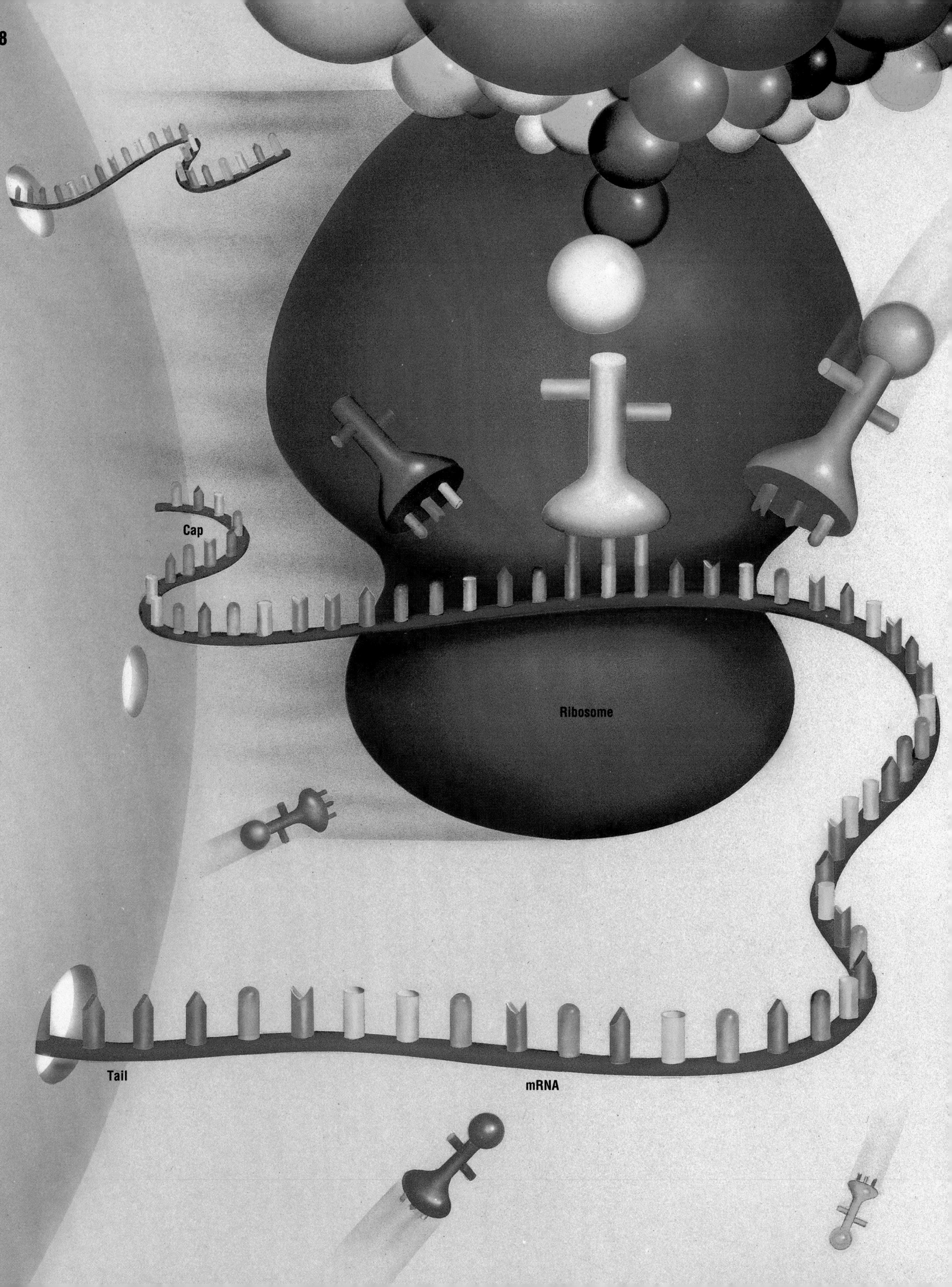

Cap
Ribosome
Tail
mRNA

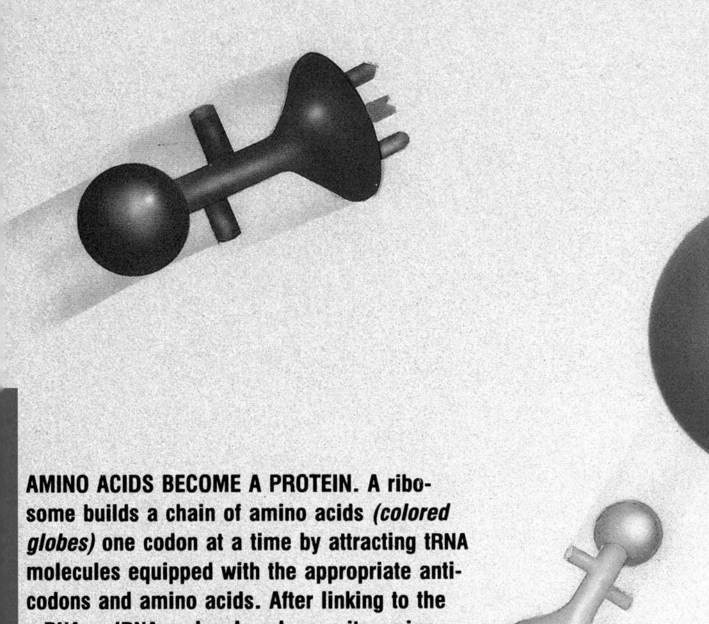

AMINO ACIDS BECOME A PROTEIN. A ribosome builds a chain of amino acids *(colored globes)* one codon at a time by attracting tRNA molecules equipped with the appropriate anticodons and amino acids. After linking to the mRNA, a tRNA molecule releases its amino acid for the ribosome to add to the growing protein molecule *(top of page)*. Then the tRNA drifts off to find another molecule of its amino acid. As shown at the bottom of this page, more than one ribosome may travel along a strand of mRNA simultaneously.

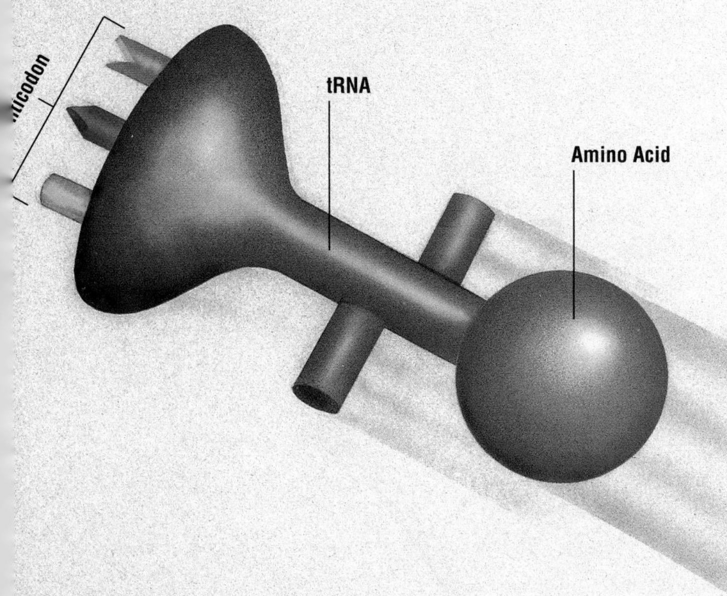

PROTEINS IN ABUNDANCE

Emerging from the nucleus, messenger RNA sets the stage for the final steps in the production of a protein. Immediately, a structure in the cytoplasm called a ribosome jumps onto the mRNA at the cap and begins the work of translating nucleotide codons into amino acids and joining them together into a protein molecule. Codon by codon, construction advances toward the tail, with forces among amino acids combining to bend each protein molecule into its own characteristic shape, without which it could not perform its assigned task.

As illustrated at left, the agents for protein building are molecules of yet another kind of RNA, called transfer RNA (tRNA). Every tRNA molecule has a trio of nucleotides called an anticodon that fits only one type of codon in the mRNA molecule. This arrangement assures that amino acids are hitched together in the correct order to yield the protein called for by the gene, now doubly decoded, that resides in the DNA.

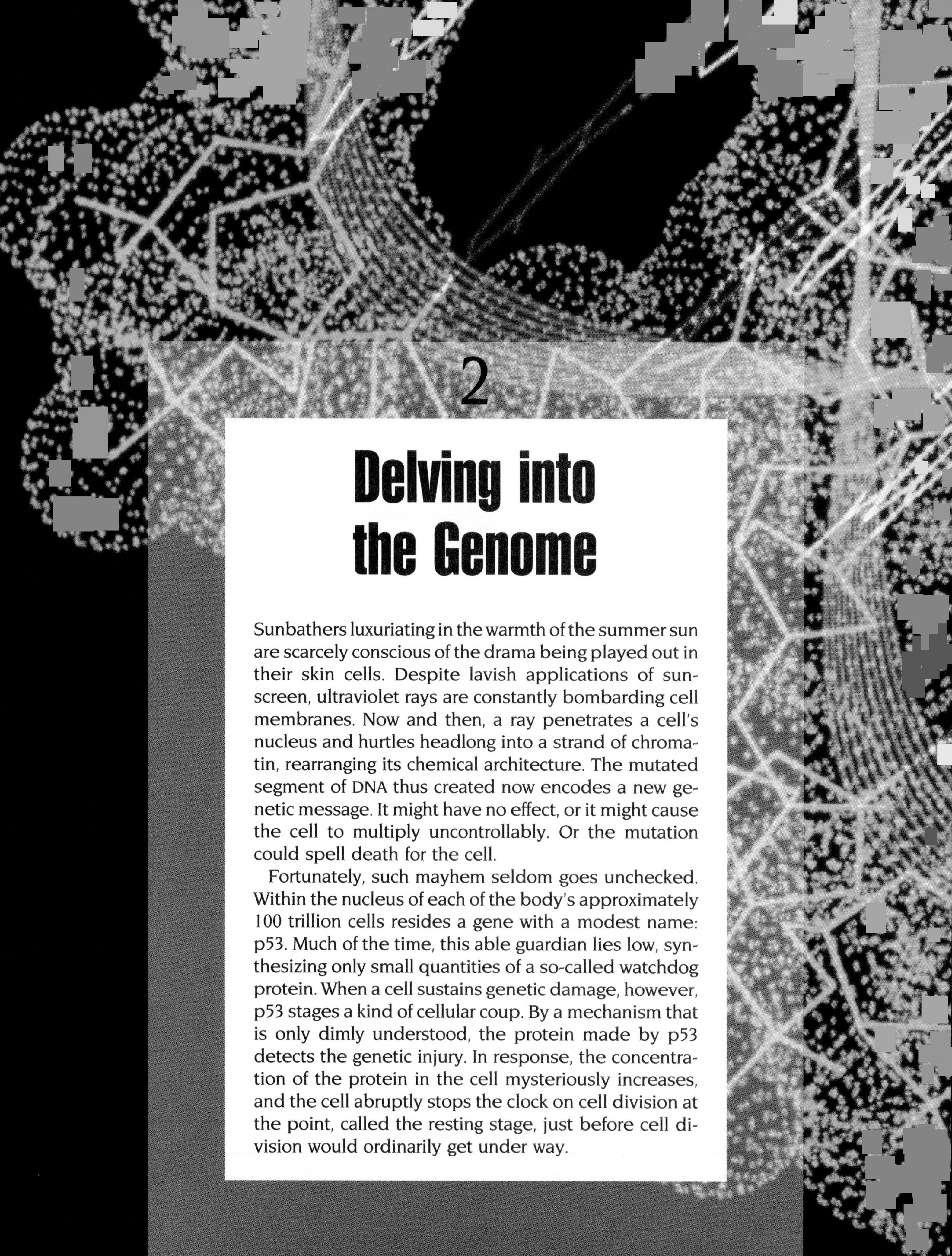

2

Delving into the Genome

Sunbathers luxuriating in the warmth of the summer sun are scarcely conscious of the drama being played out in their skin cells. Despite lavish applications of sunscreen, ultraviolet rays are constantly bombarding cell membranes. Now and then, a ray penetrates a cell's nucleus and hurtles headlong into a strand of chromatin, rearranging its chemical architecture. The mutated segment of DNA thus created now encodes a new genetic message. It might have no effect, or it might cause the cell to multiply uncontrollably. Or the mutation could spell death for the cell.

Fortunately, such mayhem seldom goes unchecked. Within the nucleus of each of the body's approximately 100 trillion cells resides a gene with a modest name: p53. Much of the time, this able guardian lies low, synthesizing only small quantities of a so-called watchdog protein. When a cell sustains genetic damage, however, p53 stages a kind of cellular coup. By a mechanism that is only dimly understood, the protein made by p53 detects the genetic injury. In response, the concentration of the protein in the cell mysteriously increases, and the cell abruptly stops the clock on cell division at the point, called the resting stage, just before cell division would ordinarily get under way.

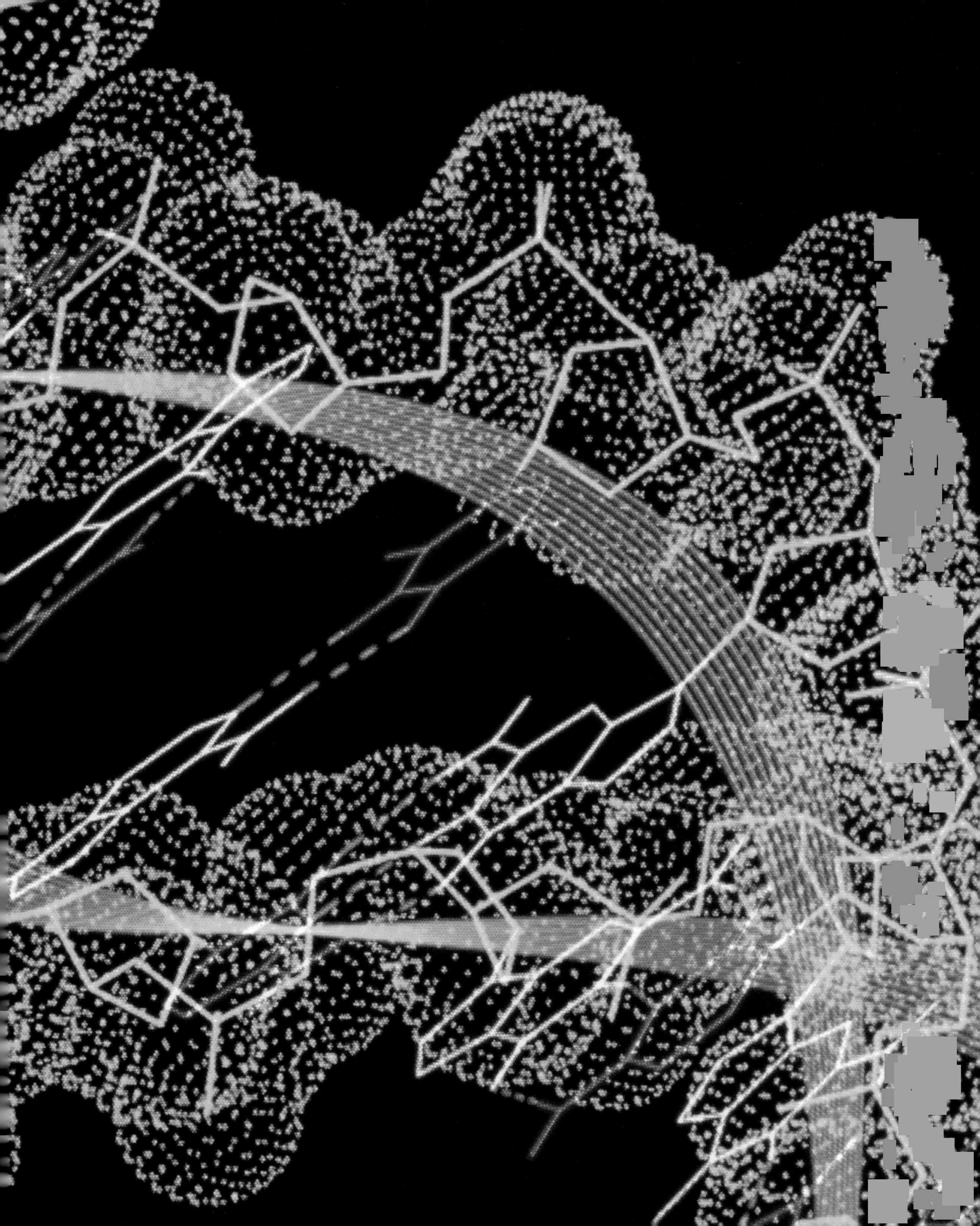

PRECEDING PAGE: Twisting blue ribbons superimposed on a computer-generated model of DNA accent the twisted ladder structure of the double-helix molecule. The sides of the ladder are colored green; the rungs, yellow and red.

Whipped into action by this comprehensive work stoppage, a crew of repair enzymes descends on the mutated chromatin. One of the enzymes surveys the damage, then another snips out the garbled segment and an "editor" enzyme stitches the correct sequence of nucleotides into the gap.

The watchdog protein, sensing that the cell's genetic house is once again in order, gradually loosens its grip on the process it has brought to a halt. Soon the cell will return to the business of reproducing itself.

Like p53, each of the body's estimated 50,000 to 100,000 genes carries the formula for manufacturing one of the equally wide variety of proteins that compose the human body. Skin and hair are made of protein, as are bones, muscles, and organs. Hormones guiding growth and development are proteins, too. The protein hemoglobin, which carries oxygen throughout the body, is found in red blood cells. Enzymes are proteins that digest food and break up other complex molecules into simpler ones. Proteins called antibodies fight disease. Even mental functions such as understanding and feeling may be regulated by proteinlike substances called neurotransmitters. In the words of noted biochemist Russell Doolittle, "Your genes supply the information, but you are your proteins."

As shown on pages 44-47, the genetic software used to create proteins is written in a code made up of DNA's four chemical bases—adenine, guanine, cytosine, and thymine. The bases are strung together in three-base sequences, or words—AGC or CTA, for example—with one or more words corresponding to one of 20 possible amino acids, the so-called building blocks of life. Messenger RNA, derived from DNA, carries the

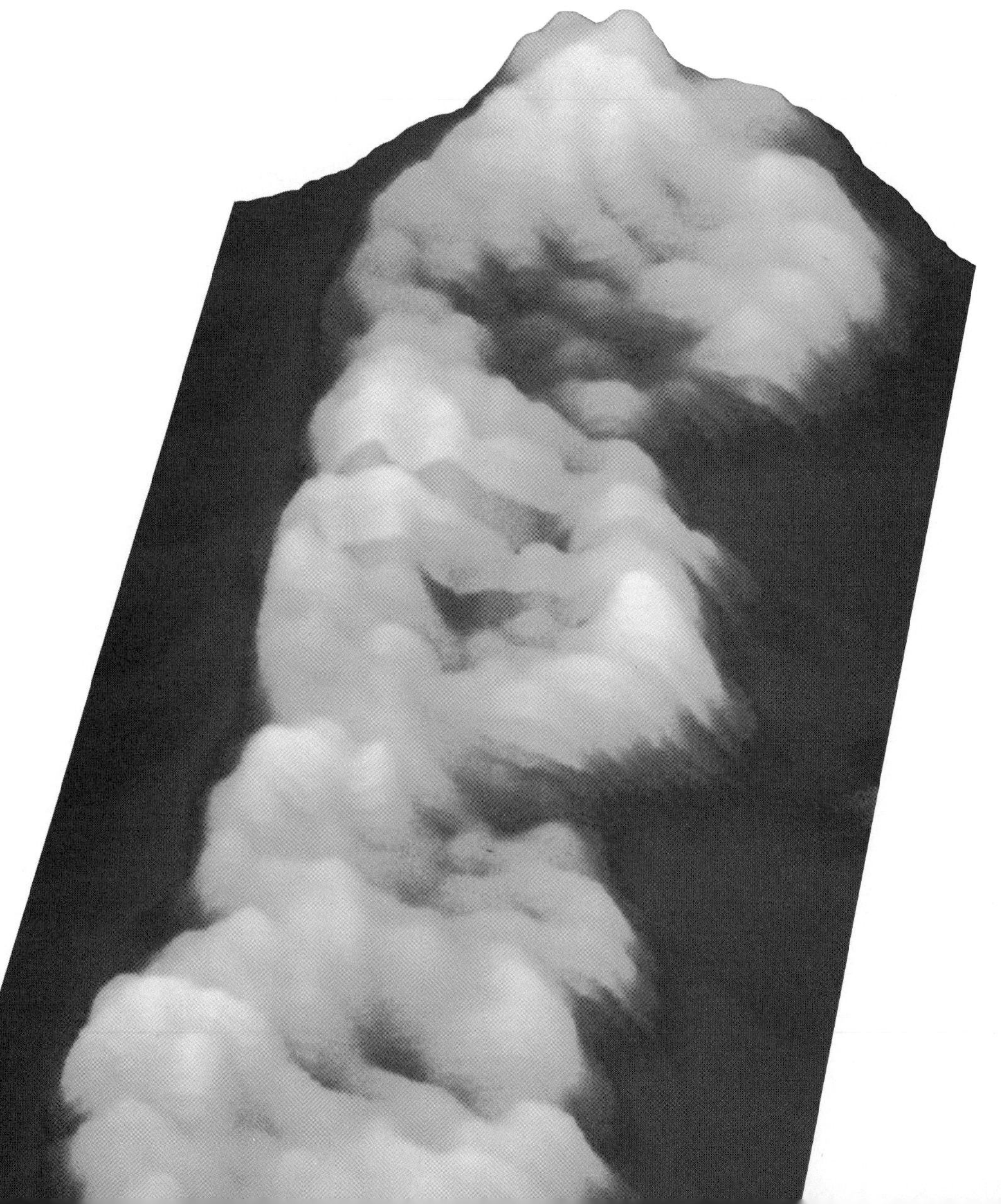

code of each gene's amino acid ciphers to molecular assembly plants outside the cell nucleus, where the information is used in the synthesis of a particular protein. Some simple proteins consist of as few as 50 amino acid molecules; the most intricate have a thousand or more.

During fabrication of even the most modest protein, complex electrical and chemical forces fold it, origamilike, into a unique shape. Some resemble corkscrews and coiled loops, while others take the form of wavy, ribbonesque sheets. Proteins perform in the many ways they do in the body because of these contours, which serve in part as docking sites for other molecules and even for genes themselves. A single error in the sequence of amino acids can corrupt the shape, turning the protein into an alien molecule that may cause a great deal of harm.

The universe of proteins divides into two broad categories, produced by two types of genes. Structural genes make the workhorse proteins that keep the body going. In cells of the stomach lining, for instance, structural genes synthesize digestive enzymes, which lock onto food molecules and chemically unbolt them into their amino acid components. Hunter-killer packs of antibody proteins ambush foreign invaders in the bloodstream and annihilate them.

Regulatory genes, the second type, manufacture proteins that control when, where, and how much of other proteins are produced. In effect, regulatory proteins act as switches to turn structural genes on or off depending on the body's needs. One type of regulatory gene, for example, makes proteins that mobilize antibody-producing structural genes, depending on what kind of germ has invaded the body. P53 manufactures a protein that halts progress toward cell division when the formula for any other protein has been mangled. To exert this authority, regulatory proteins often affix themselves directly to the gene. One class of regulators actually clasps the DNA helix in a scissors grip the way wrestlers wrap their legs around opponents' torsos.

Despite cells' heroic efforts to prevent genetic errors and to correct the very few that occur, the system is imperfect. Changes in the DNA can be so massive that the helix simply cannot be fixed. A smaller error may occur in a place critical to the cell's survival, or the damage may be so subtle that it goes undetected. A gene that happens to move from one chromosome to another (*page* 75), for example, is flawless except for its altered location. In its new spot, it may seize control of a nearby structural gene, stopping that gene from making a vital protein or, conversely, activating it in a cell that should not be manufacturing the substance.

Rarely are such mutations beneficial. They are far more likely to cause debilitating and even fatal conditions—cancer, for instance—in the individual whose DNA is rearranged. Perhaps worse, some mutations can plague generation after generation as the altered gene passes from grandparents to parents to children and beyond.

Until recently, the best that could be done, and then only in a very few genetic disorders, was perhaps to

Individual atoms appear as tiny, fuzzy balls in this image of DNA magnified more than 10 million times. Made with a scanning tunneling microscope—a device that forms a computer image from the undulations of a finely pointed probe as it follows the contours of the molecule a few atom diameters above its surface—the micrograph shows almost three turns of the molecule's double-helix structure.

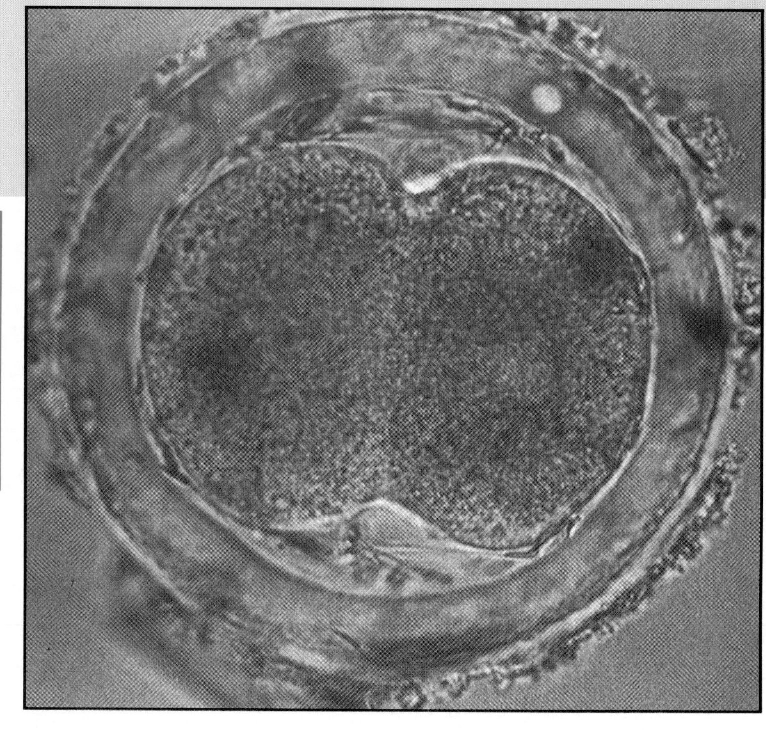

During the two days after fertilization, a human embryo first has two cells, then four, then eight, all enclosed temporarily in a membrane. At this early stage, the cells are totipotent. That is, they have the potential to become any kind of tissue in the body. But after only one more cell division, when the embryo contains 16 cells, they begin to differentiate, or take on the unique appearance and function of a particular body tissue *(overleaf).*

supply a missing protein, such as insulin in the case of diabetes. Other disorders, like the lung disease cystic fibrosis, were unassailable beyond temporary relief of symptoms. Since the late 1970s, however, scientists have begun to make progress toward truly ingenious ways of attacking some of these ailments at the source with new, functioning genes that take over for damaged, inoperative ones. Although methods are unproven and results tentative, it is now possible to foresee a day when many intractable diseases of genetic origin could succumb to genetic remedies as reliably as many disease-causing bacteria yield to assault by antibiotics.

Genes are not free agents. Were they allowed to produce their proteins willy-nilly, chaos would result. Life would not be possible.

Scientists have long noted that almost every cell in the body contains in its nucleus the complete, unabridged edition of the human genome. Yet on average, a mature cell expresses only a small percentage of the genetic message it carries. That is, it manufactures only a tiny portion of the proteins for which there are recipes in the nucleus.

The sorting out of which genes in which cells will produce proteins begins very early in development. From conception onward, the genes in each dividing cell are turned on and off in a carefully prescribed order that steers each cell toward a distinct developmental destiny. Called differentiation, the process dictates that some cells become skin, others muscle, and still others blood, kidney tissue, or neurons.

Only a few thousand genes are active throughout the entire body, operating in virtually every fully developed cell. For instance, p53 is at work in all cells, guarding against the perpetuation of mutations that might occur if cells were to replicate damaged DNA. Other widely active genes are responsible for products such as cytochrome C, a protein involved with conversion of the sugar glucose into energy for cell operations, and DNA polymerase, an enzyme crucial to the replication of DNA.

Much more common are genes that manufacture proteins in only a few types of cells. This division of labor can be exquisitely detailed. For ex-

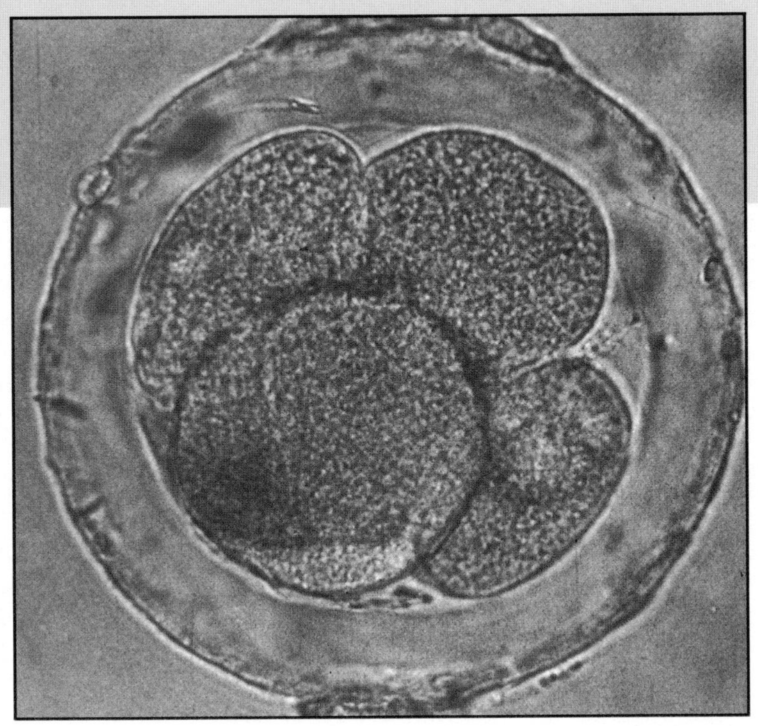

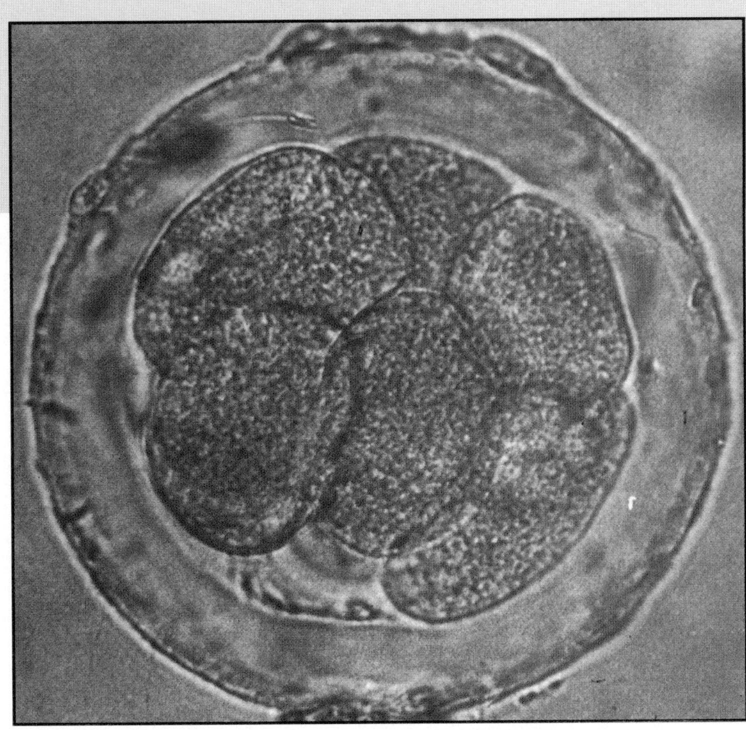

ample, a nerve cell, which produces neurotransmitters crucial to relaying nerve impulses to and from the brain as well as within it, rarely turns out more than one or two of the 200 or so such agents thought to exist.

While scientists have long proposed the theory that some kind of master control system was at work orchestrating this developmental score, until recently they had little direct evidence to support their hypothesis. Then, in the mid-1980s, molecular biologists discovered a class of regulatory gene that they believe is responsible for mapping out the body's head-to-toe architecture when the human organism is little more than a bubble of cells. These master regulators, known as homeotic genes, are thought not only to stipulate what position cells will occupy in the developing organism—becoming part of the head, a foot, or something in between—but also to dictate which proteins each cell is to make, thereby establishing the cell's identity: bone or brain tissue, heart muscle, or light-sensitive rods and cones in the eye's retina.

The discovery of master regulators came about quite unexpectedly, through the study of fruit-fly embryos. For close to 50 years, embryologists seeking clues about the genetic basis of development had been probing the genome of this unimposing insect for evidence of specific linkages between genes and body parts. Fruit flies are ideal for such studies because they offer two advantages for genetic researchers: a short life cycle, enabling the researchers to evaluate experimental results quickly; and a genome of only about 5,000 genes, greatly simplifying the hunt for a specific gene.

During experiments with fruit flies beginning in the 1950s, researchers found that, by aiming tiny bursts of x-rays at male flies, they could cause mutations in the fly offspring that would seem right at home in a science-fiction thriller. Wings sprouted where eyes should be. Legs supplanted antennae, and so on.

When experimenters compared the chromosomes of these miniature

These four diverse tissues are all descended from the embryonic cells shown on pages 54-55. Clockwise from left, they are striated muscle, the body's principal muscle cell; skin cells; calcium-rich bone cells seemingly stacked like planks; and a tangle of nerve cells.

monsters with those of normal flies, they made a wholly unexpected discovery: Each of the intriguing transformations had resulted from damage to just one gene, not to the hundreds of genes necessarily involved in the shaping and correct placement of legs and other complex body parts.

It stood to reason that the affected gene must control a constellation of subordinate genes. But to turn theory into fact, investigators would have to isolate the critical gene and establish the sequence of base pairs within it. Only by doing so could scientists explain how this influential member regulated all the other genes involved in development.

For nearly 20 years to come, however, there would be no practical way of singling out an individual gene from among the thousands found in higher organisms like the fruit fly. Then, toward the end of the 1970s, new techniques were invented for creating probes to isolate individual genes and determining the sequence of base pairs within them.

To fashion such a probe, a type of genetic material called messenger RNA (mRNA)—which serves as the template for manufacturing proteins (*pages* 46-49)—is taken from the cytoplasm of normal fruit-fly cells. When mixed with enzymes in a soup of nucleotides, the mRNA serves as a model for assembling strands of complementary DNA, or cDNA, that are exact replicas of a small part of one side of the double helix inside the cell nucleus.

In a test tube containing strands of DNA, the probe unerringly seeks out and attaches to the unique sequence of nucleotides belonging to the gene in question. Tagged in advance with a bit of radioactivity or fluorescent dye to make it stand out as a marker, the

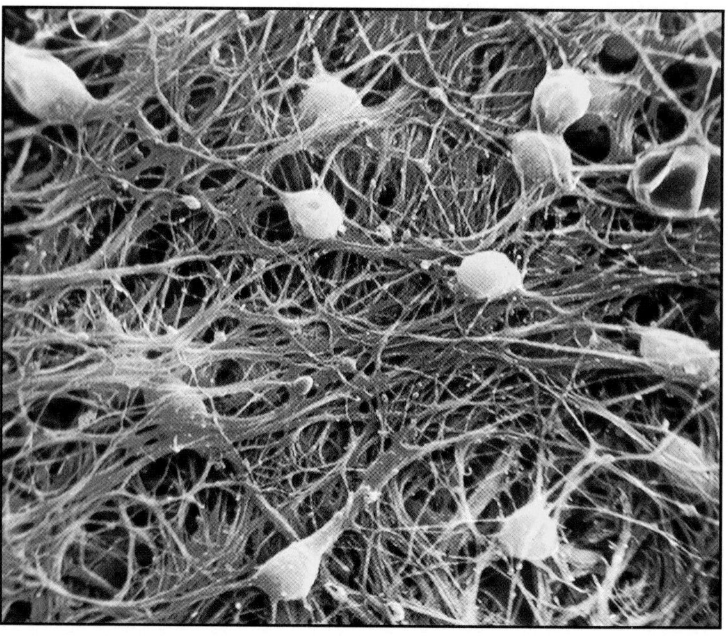

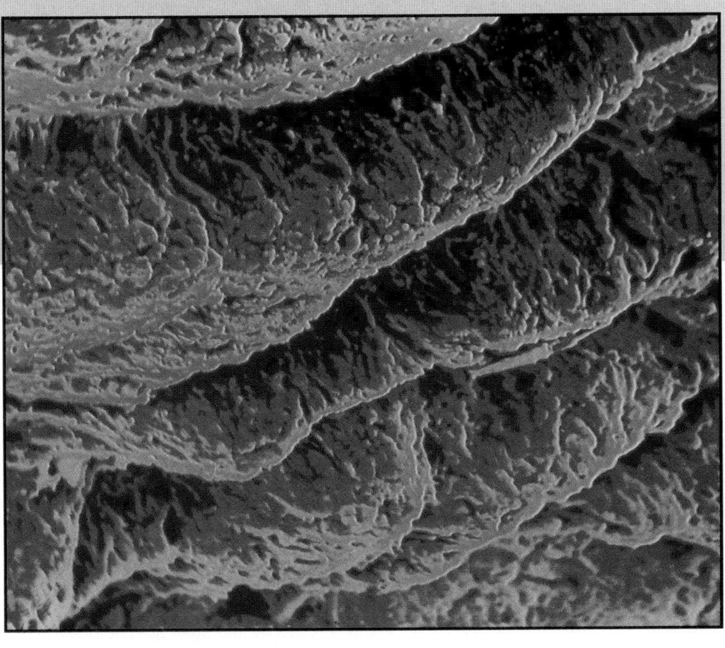

probe serves as a beacon to the gene's location.

In the early 1980s scientists in Switzerland and the United States made a probe tailored to latch onto fruit-fly DNA at one place only: the homeotic gene. As expected, the probe made straight for the gene targeted by the investigators. But much to the researchers' astonishment, the supposedly unique probe flagged identical nucleotide sequences elsewhere in the fruit fly's DNA. Soon more than a dozen fruit-fly homeotic genes—all by definition involved in body-part formation—were found to contain the same sequence of 180 base pairs.

This strip of nucleotides—hallmark of a master regulator gene—was dubbed the homeobox. Genes carrying it are one of the fruit fly's frontline developmental elite. They help to establish the schedule of events as the organism develops from a fertilized egg, as well as which parts end up where in the adult insect.

Discovery of the homeobox spurred a research frenzy that one geneticist has wryly named "homeomadness." Biologists in western Europe and across the United States loosed swarms of radioactively tagged homeobox probes into the genomes of jellyfish, worms, frogs, mice, and humans. Without exception, each of the species investigated was found to have homeotic genes that control its early development.

But there is more. Even though the total sequence of base pairs in the homeobox varies considerably among different species, the proteins from genes containing the homeobox are nearly identical. Certain homeotic proteins produced by frogs and fruit flies, for instance, match each other, amino acid for amino acid, in 55 positions of the 60-amino-acid chain.

And not only that. In 1990 William McGinnis of Yale University in New Haven, Connecticut, announced the results of a radical experiment. He and his assistants replaced the homeobox on the master regulator gene responsible for head development in a fruit-fly embryo with a homeobox taken from a human cell. The doctored fruit fly grew up with a head virtually identical to one whose development was governed by the insect's own homeotic genes, even though the fruit fly evolved some half a billion years before human beings came on the scene.

Edward Lewis, a geneticist at the California Institute of Technology in Pasadena, speculates that the homeobox has been carefully preserved over time simply because the body plans of most life forms are so alike. "Humans and flies are very similar in organization," Lewis points out. "They both have a head, thorax, and abdomen, because that seems to be the most sensible and efficient form. Why would evolution invent all these things over again?"

Scientists do not fully understand yet how homeotic proteins interact with other genes to dictate the specialization of cells into so many different kinds, from neurons in the brain to a strip of cartilage in the knee. But a clue to the works appears in three-dimensional views of homeotic proteins produced by a computerized imaging technique known as x-ray crystallography.

These protein renderings show that one small segment of the molecule is shaped like a helix and doubled over, hairpin fashion. By clamping down with the two prongs of the hairpin, homeotic proteins anchor themselves to the gene. During development, the protein either silences the gene or calls on it to produce proteins of its own that play defining roles in deciding each cell's future contribution.

The first thing a developing cell learns from its homeobox overseers is where it belongs in the growing organism. Studies of fertilized fruit-fly eggs at the Max Planck Institute in Germany suggest a means by which such information is communicated among the cells.

Well before fertilization, nurse cells that nourish the fruit-fly egg inject maternal RNA into the egg at one end. After fertilization, the RNA triggers protein production by a homeotic gene named bicoid. The bicoid protein diffuses through the developing embryo in ever-decreasing concentration. Cells in the region where the bicoid protein is most abundant become part of the embryo's head; those at the other end of the embryo, with the smallest quantity of the protein, become the tail. Other homeotic genes later govern the further specialization of cells as parts of the fly's body structure—eight segments in the abdomen, three in the thorax, and three in the head.

Homeotic proteins need assistance in steering immature cells to their ultimate occupations. Their accomplices are courier proteins—produced on orders from the homeotic genes—which shuttle biochemical signals between nuclei in developing cells. Since cells in an embryo migrate considerable distances during development, they are exposed to a variety of courier proteins that selectively influence their genes, prodding cells toward one function or another.

The lens cells of a frog's eye are a case in point. Early in development they are located above the frog's throat, where they first function as skin cells. But later, cells from the frog's future heart brush past the skin cells, telegraphing signals that stimulate the first of the lens-making proteins. Only after the nascent lens cells complete a journey of their own that brings them into contact with maturing retinal cells do they receive the final signals that turn them into fully formed lenses.

Scientists' knowledge of these

mechanisms is both crude and incomplete. For example, why the same signal that sparks the development of a wing in a fruit fly prompts the growth of a skin cell in a human being remains a puzzle made all the more perplexing by the close similarities between the homeoboxes of the two species. Until that mystery and others are solved, scientists are unlikely to understand the complicated specialization implied, for example, in the human brain's billions of interconnecting neurons.

Yet the study of these phenomena in fruit flies, frogs, and other creatures will pay dividends in understanding generally—if not specifically—how human beings develop. "What we are seeing is that the basic building blocks of all organisms were in place 500 million years ago when flies, worms, and humans diverged," says Gerald Rubin, a molecular geneticist at the University of California at Berkeley. "The types of switches, the types of wires, are all the same."

In the field of genetics, unraveling the mysteries of how an embryo eventually becomes an adult has few rivals for sparking awe and wonder. Differentiation only explains how cells become dedicated to the production of some proteins but not others. Of equal interest and importance is how regulatory genes cause structural genes to stop and start the manufacture of these proteins on demand, throughout life. A discovery made in 1979 by Richard Flavell and L. H. T. van der Ploeg, two geneticists working at the University of Amsterdam, has permitted scientists a glimpse into the mechanisms by which the genome orchestrates the switching of genes on and off.

This fruit fly has two pairs of wings instead of the usual single pair because of a mutation in one of several genes that control tissue development in this and most other organisms. The mutation, provoked in the laboratory by exposing the embryonic fly to x-rays, caused a doubling in the number of wing-bearing segments of the insect's thorax.

Prologue to a Fruit Fly

Among biology's most profound mysteries is the means by which cells in a human embryo develop into specific tissues. In part because of unexpected similarities among different species in the homeotic genes that appear to regulate the process, scientists have relied on the well-studied fruit fly for enlightenment.

As explained at right, the road from fertilized egg to fruit fly begins with segmenting genes that establish the head-to-tail layout of the insect. Then homeotic genes begin the formation of internal and external features. Judging from results of experiments with mice, a process similar to the one that tells a fruit fly which way is up appears to be at work in the embryos of mammals, including humans.

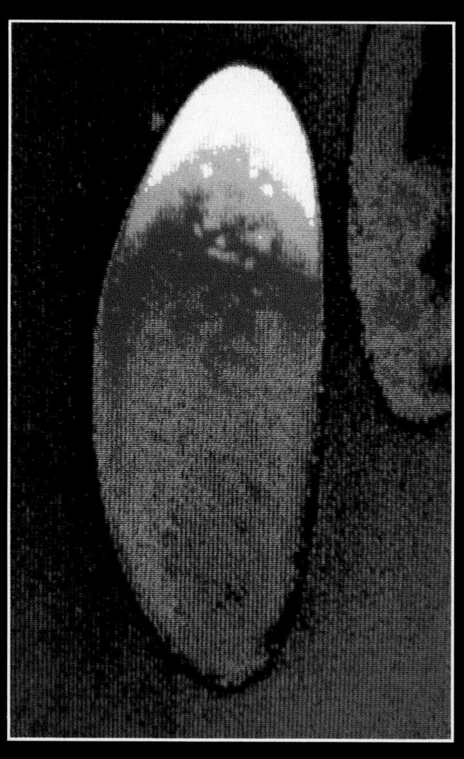

1 HOUR. Yellow, orange, and red zones in this computer-generated image of a fruit-fly embryo reveal high levels of bicoid protein, which marks the future head and thorax. At the bottom of the egg, the protein nanos *(not shown)* marks the abdomen and tail.

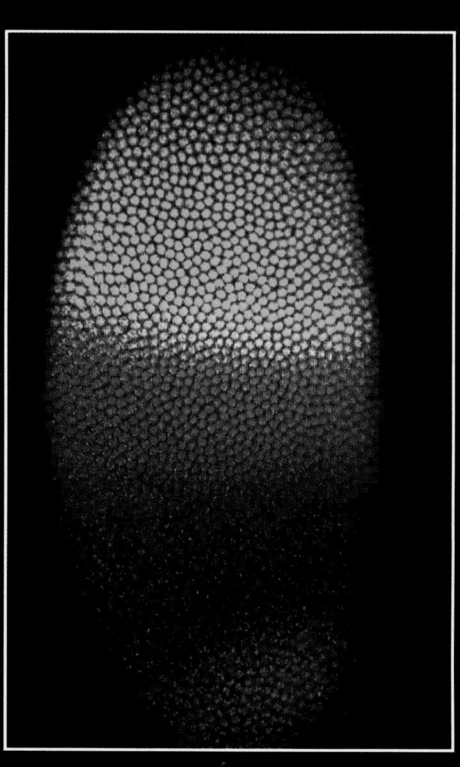

2-1/2 HOURS. Segmentation begins as gap-gene proteins divide the embryo into front *(bright green)*, middle *(red)*, and rear *(dark green)*. The yellow band represents the area where front and middle proteins overlap.

At the time Flavell and van der Ploeg were studying a well-known instance of protein regulation, in which production of a type of hemoglobin abundant in fetuses is halted and a slightly different formulation of the same oxygen-carrying protein in adults is begun.

From about the third month of a fetus's life in the womb until shortly after birth, the baby manufactures molecules of hemoglobin that are made up of two pairs of interlocking protein chains known as alpha-globin and gamma-globin. Separate genes code for each of the chains. The gamma-globin gene lies on chromosome 11, while the alpha-globin gene is found on chromosome 16. Once the two proteins are assembled, they somehow find each other and link up.

The alpha-globin gene remains active into adulthood. But for some unknown reason, the gamma-globin gene begins to curtail its activity by the time the fetus is six months along. This gene's role is gradually assumed by a neighbor on chromosome 11, the so-called beta-globin gene. By a child's first birthday, almost all of the hemoglobin molecules contain two alpha-globin chains and

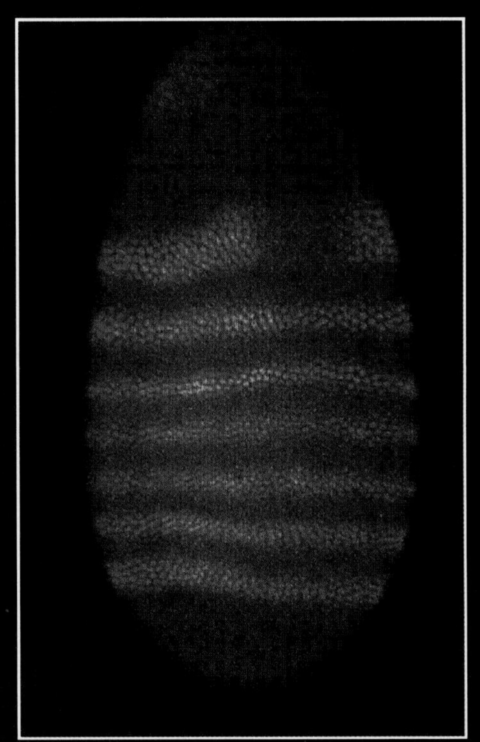

3 HOURS. Three bands become seven stripes of protein produced by the so-called hairy gene, named for its function later in development: to make the bristles that are found on adult fruit flies.

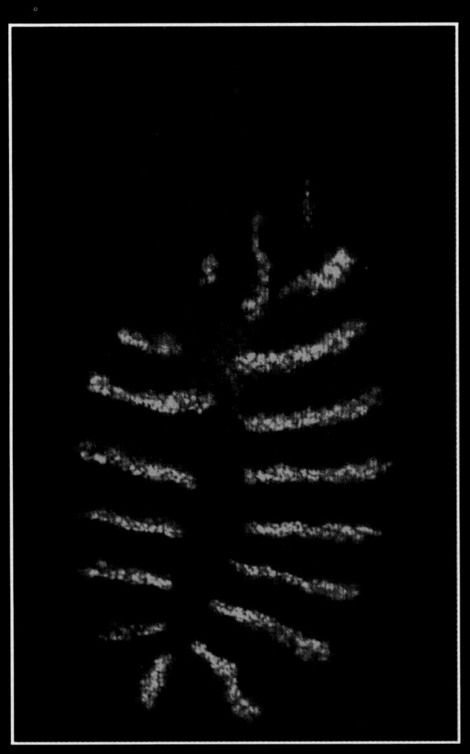

4 HOURS. The number of stripes doubles as the embryo elongates and folds near the middle. Beginning at the top of the left side and going counterclockwise, the first three stripes mark the head, the next three the thorax, and the rest the abdomen.

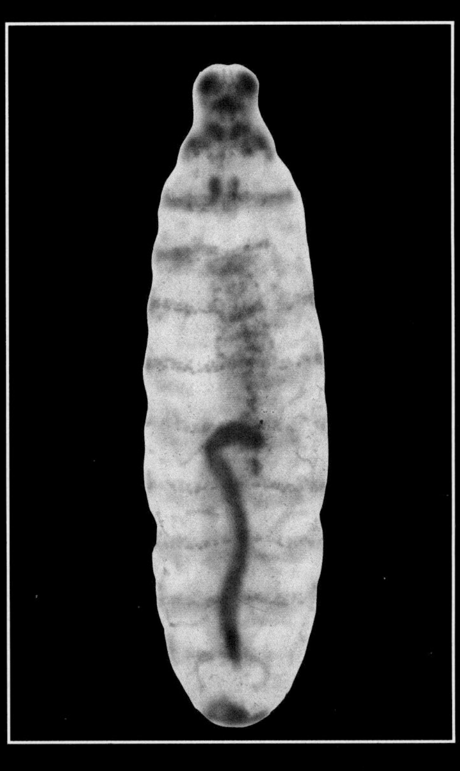

24 HOURS. Segments are clearly visible as recognizable structures appear in the hatched larva. Spots at the top belong to the head. Part of the gut shows up as a dark walking-stick shape, while just above it faintly emerges the fruit-fly equivalent of a spinal cord.

two beta-globin chains. Few gamma-globins are to be found.

Flavell and van der Ploeg subjected gamma-globin genes from fetuses and adults to detailed chemical analysis. When the gene is inactive, as in adults, the investigators found, a simple hydrogen-and-carbon molecule known as a methyl group attaches itself in some of the crannies on the outside of the double helix. Conversely, the active form of the gamma-globin gene found in fetuses was noticeably free of methyl groups. The difference led the Dutch researchers to speculate that methyl molecules might be the "switches" that turn genes on and off.

A serendipitous bit of data to support this theory soon came from unrelated laboratory experiments in which a potent anticancer drug caused cells in adult tissue cultures to begin producing fetal gamma-globin. Just as Flavell and van der Ploeg might have predicted, examination of the gene responsible for producing this protein showed it to be almost free of methyl groups. In recent years, other inactive structural genes have been found to include methyl groups, further evidence that these simple additions to the DNA structure can be important to the body's gene-regulating scheme.

Some scientists wonder whether such details might not hold explanations not only for comparatively clear-cut genetic disorders such as various forms of anemia, but also for phenomena as commonplace and seemingly inevitable as growing old. A cell's accidental loss of the ability to attach methyl groups to the appropriate genes or to disperse them, goes this nascent theory, could activate formerly silent genes or turn off those whose proteins are much needed.

Should this hypothesis someday be proved correct, one of the consequences might be to show that aging results from the cumulative effects of such corruption over decades of DNA replication. "The biological identity crisis that we define as old age," suggests geneticist Steve Jones of University College, London, could happen "when the genetic message has become so degenerate that the instructions no longer make sense."

Considering the countless number of times the genome is copied in a lifetime, the human body's durability should perhaps seem more of a miracle than it usually does. The genetic instructions packed into the chromosomes of each cell are some three billion characters long. Even if genetic defects from environmental damage or coding errors were as scarce as one in every million bases, 3,000 mistakes would be carried forward during every cell division, which scientists estimate occurs a million-billion times before birth alone. Because of p53 and a cadre of other watchdog genes, the frequency with which chromosome damage and copying blunders escape correction is closer to one in 10 billion. Even so low an error rate, however, still has the potential over a life span to transform the intricate code of life into elaborate gibberish.

Although geneticists may someday discover the fountain of youth hidden within the human genome, they expect to make progress much sooner with less sweeping infirmities. There are plenty of candidates, from diabetes and various kinds of anemia to disorders of the immune system and even many forms of cancer.

Since the dawn of the genetic age in the early 1960s, scientists have dreamed of using their knowledge to engineer cures for such genetic killers. The most elegant remedy—the simple replacement of the mutant gene—was, back then at least, no more than a fantasy. So rather than target genes, researchers initially focused their energies on a more accessible quarry: the protein. If they could "build" flawless proteins to take the place of faulty ones, they could perhaps produce a cure or at least alleviate some disease symptoms.

Scientists' first tentative attempts at protein manufacturing met with resounding success. By the early 1970s, they had synthesized insulin, the protein (in this case a hormone) responsible for shuttling glucose from the bloodstream to cell interiors, where it serves as a source of energy.

The pathbreaking technique was relatively straightforward, as biochemical processes go: A human "protein engineer" began with a wand made of resinous material and a vat with a solution containing the first amino acid in insulin's 51-link chain of such molecules. Also mixed into the solution were other chemical boosters that induced the amino acid to latch onto the wand, which was then transferred to a vat filled with the second amino acid—and so on for all 51. The ultimate result, after some 20 days of mostly waiting and watching, was a substantial quantity of insulin, which was stored for future use.

But what works for simple proteins like insulin turned out to be utterly useless for baroque molecules such as hemoglobin. Even though researchers had deciphered the sequence and the three-dimensional structure of hemoglobin's 574 amino acids, constructing such a large molecule by hand became a project so unwieldy as to be impossible.

There is another way to construct molecules, however: a gene-splicing technique known as recombinant DNA. Developed even as insulin was being assembled amino-acid-by-amino-acid in the laboratory, the procedure recruits living cells—and they need not even be human ones—as an army of protein synthesizers.

The process works like this. First, lab technicians isolate from human cells the gene for the desired protein, then use an enzyme to excise it from its chromosome. Next, they mix the liberated genes with snippets of DNA extracted from bacteria. They often use a common one named *Escherichia coli.* (Found in human intestines, the organism usually goes by the abbreviated name E. *coli.*) An enzyme added to the preparation links the human and bacterial DNA, creating a hybrid—or recombinant—molecule. This hybrid is inserted into a live bacterium, which is then cultured.

"The bacterium is stupid," as endocrinologist Ron Rosenfeld, a trailblazer in protein studies at Stanford University in California has put it. "It accepts the new gene as its own.

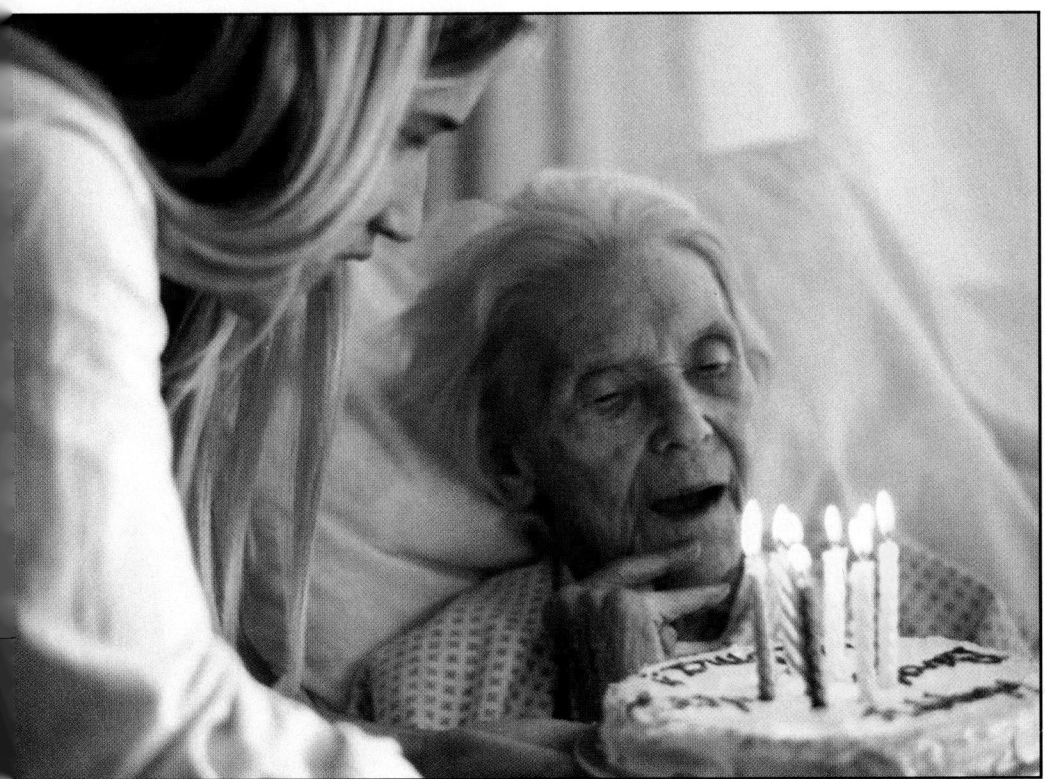

At a Santa Monica nursing home, a young woman presents a birthday cake to an elderly patient. Geneticists theorize that wrinkles and white hair, as well as other signs of aging, come in large measure from the unavoidable accumulation of mutations in a person's genes during a lifetime of cell divisions.

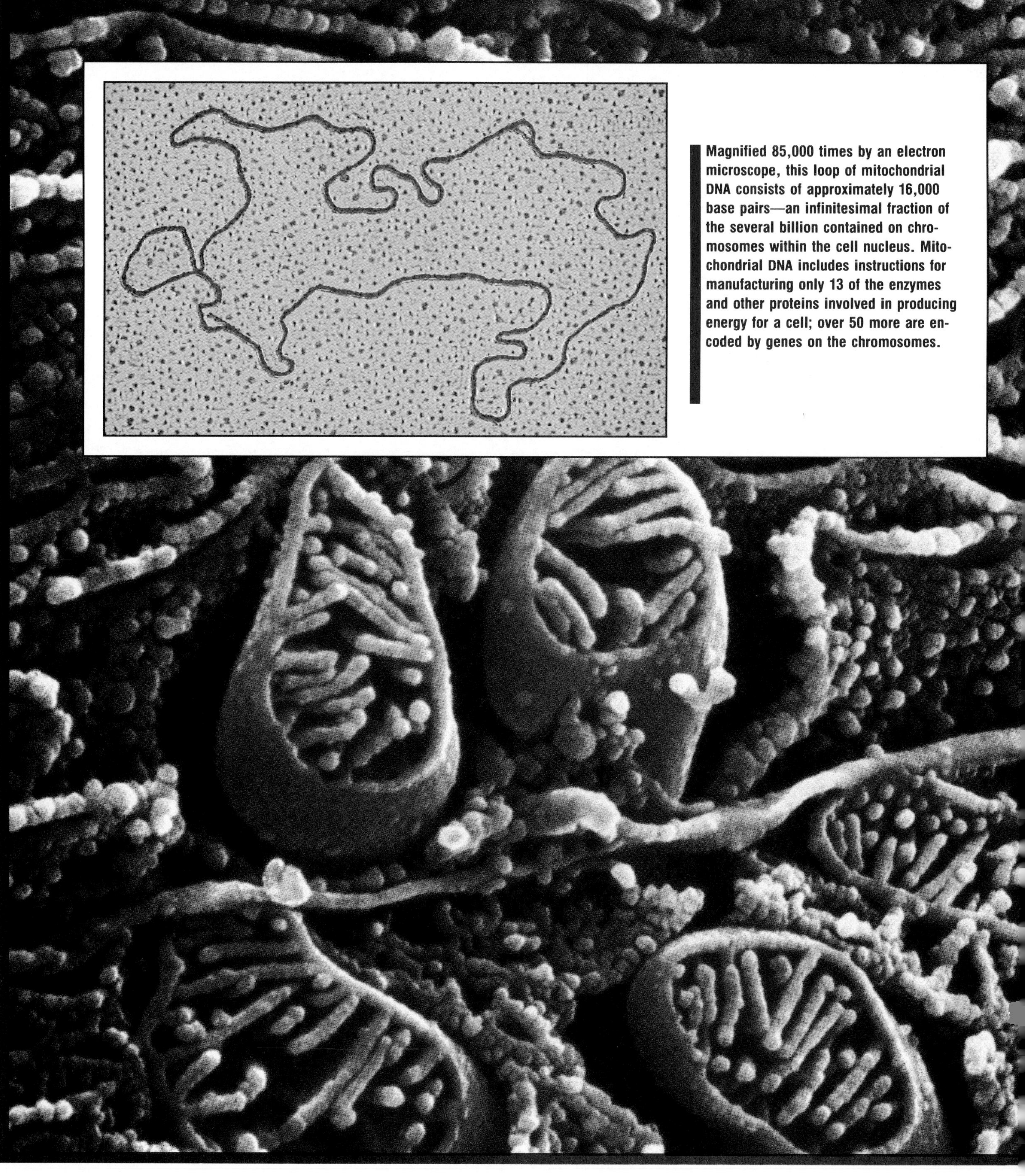

Magnified 85,000 times by an electron microscope, this loop of mitochondrial DNA consists of approximately 16,000 base pairs—an infinitesimal fraction of the several billion contained on chromosomes within the cell nucleus. Mitochondrial DNA includes instructions for manufacturing only 13 of the enzymes and other proteins involved in producing energy for a cell; over 50 more are encoded by genes on the chromosomes.

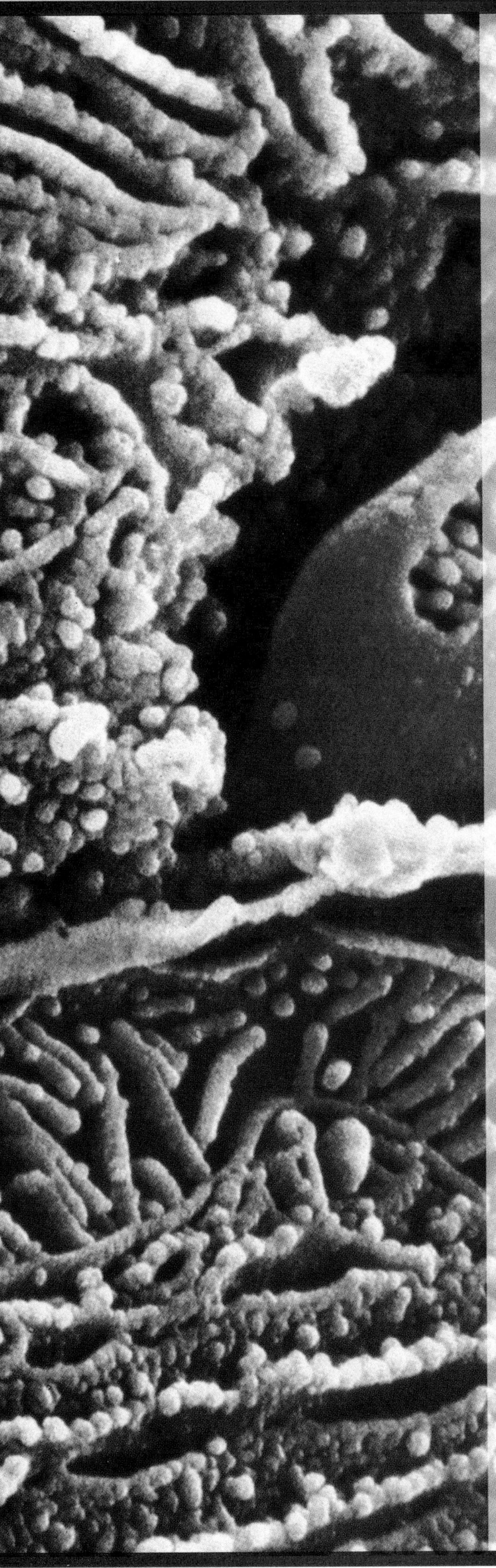

Coding for Energy: A Maverick Brand of DNA

Individual cells, like whole organisms, perform certain tasks and need energy to do so. Within the cytoplasm of cells, tiny structures called mitochondria supply that power by employing special enzymes to break down nutrients and convert their energy into a form the cells can use. In the course of investigating this process, scientists have found, to their surprise, that the DNA coding for some of these enzymes resides within the mitochondria themselves—not on the chromosomes in the nucleus with all the other genetic instructions that control a cell's form and function.

Mitochondria and their DNA have other strange features as well. For example, they are the only parts of most cells that duplicate themselves independently, separate from the process of cell division. Also, strands of mitochondrial DNA form rings—unlike the threads of chromosomal DNA in the nucleus, but exactly like the DNA in bacteria. Indeed, portions of the DNA sequence of some mitochondrial genes are very similar to the coding found in certain kinds of bacteria.

All this has led researchers to conclude that mitochondria were once independent bacterial cells that somehow became incorporated into a larger but metabolically less efficient host cell some billion and a half years ago. Both cell types benefited as a result. The bacteria apparently made use of compounds broken down by the host cell for their own energy needs, producing excess energy that was in turn employed by the host. Over time, the two cell types became so integrated that they evolved into a single complex cell. The former bacterial components became increasingly specialized, retaining only those functions necessary for energy production and thus developing into an essential part of the cellular machinery.

Nestled within the cytoplasm of a human liver cell, several mitochondria *(pink bodies)*—each less than a thousandth of a millimeter across in actual size—reveal the convoluted folds of their inner membranes. These folds are repositories for the molecules that take part in converting nutrients into energy. A single liver cell can contain as many as 2,000 separate mitochondria.

When the gene instructs the E. *coli* to start creating the protein, the E. *coli* complies. In effect, the bacterium becomes a little factory for whatever we want to make."

Within 15 hours, the recombinant cell generates more than a billion identical copies of itself. These clones are nurtured in vats, where they continue to multiply and to crank out enormous quantities of the protein. When they have made enough, the bacteria are killed and the protein purified. Today, commercial biotech labs use recombinant DNA technology to produce not only simple proteins such as insulin, but also complex ones such as hemoglobin and tissue plasminogen activator, the body's own blood-clot solvent.

There is incentive aplenty to synthesize hemoglobin. Diseases caused by mutations in the globin genes are the most common genetic illnesses in the world, no doubt because globin genes are among the busiest known. Over a period of four months, the body replaces its entire supply of red blood cells. This is a tremendous undertaking. Each of the six or so quarts of liquid coursing through the blood vessels contains more than five trillion red blood cells, each packed with 280 million hemoglobin molecules. Manufacturing just one of these molecules requires transcribing the 1,700 base pairs necessary to assemble beta-globin and alpha-globin from amino acids.

One of the most intractable hemoglobin diseases is sickle cell anemia, an often fatal condition characterized by bouts of fever and searing pain in the bones, joints, and abdomen. The disorder is caused by a single base-pair mutation in the beta-globin gene. This infinitesimal error causes hemoglobin's amino acids to fold up slightly askew, so that the finished molecule has on its surface both a protruding knob and a receptacle into which other defective hemoglobin molecules can plug themselves.

As a result, interlocking chains of the malformed protein assemble inside their host cells, sometimes warping them into a sickle shape. The distorted cells pile up in narrow capillaries and impede blood flow. Internal organs soon become oxygen starved and the body begins to fail.

Unfortunately, a treatment for sickle cell anemia and other hemoglobin disorders has not followed from the ability to make bacteria produce hemoglobin. The volumes of the protein required for therapeutic use, the cost of producing it, and the fact that it cannot be placed inside red blood cells where it normally functions, but only injected into the blood plasma—where it is only marginally effective—have thwarted its use in patients. Yet nowhere is the admonishment to try, try again taken more to heart than among medical researchers. If bacteria are ultimately impractical as hemoglobin factories, why not put the reengineered gene directly into human bone marrow cells that produce the protein, in effect repairing the body's defective beta-globin gene? In 1979 a 25-year-old graduate student at Stanford named Richard Mulligan took the first step toward that goal with a pioneering experiment.

Borrowing a leaf from the recombinant DNA manual, Mulligan spliced a rabbit gene for hemoglobin into some bacterial DNA. But instead of putting his rabbit-hemoglobin factory into E. *coli*, Mulligan slipped it into several million copies of a virus that infects monkeys, having first destroyed the virus's ability to order itself reproduced by its host but leaving intact its mechanisms for instructing the host to make proteins.

Mulligan then took the genetically altered viruses and placed them in a dish containing a culture of cells from a monkey's kidney. His guess was that the genetically defanged virus would tell the kidney cells to make hemoglobin. He was right.

With this remarkable gene-transfer

experiment, Mulligan had converted an ordinary virus—nature's wiliest invader of cells—into a biological "truck" for conveying genes into living cells (*pages* 98-99). Suddenly, the medical world seemed on the threshold of having a powerful new genetic tool at its disposal.

Mulligan's procedure, however, was a long way from becoming a genetic treatment for disease. Although the monkey virus infected all the kidney cells exposed to it and harnessed them to the production of hemoglobin, the microbe could not infect human cells. Furthermore, the kidney-cell hemoglobin was of no value in treating anemia, because the protein could not move from those cells into red blood cells. Thus Mulligan needed to find a different truck, one that would work not only in humans but in their stem cells, the body's lifelong source of blood cells both red and white. Normal hemoglobin genes toiling in stem cells would, in theory, give new red cells an adequate ration of the protein in its unmutated form.

In a different class of virus, called retroviruses, Mulligan saw the potential to infect stem cells in the way necessary to get normal hemoglobin into red blood cells. Retroviruses reproduce somewhat differently from others. Whereas DNA from a monkey virus merely takes up residence in an infected cell's protoplasm, a version of the genetic program from a retrovirus, called the provirus, shoulders its way into one or more of the cell's chromosomes (*page* 72).

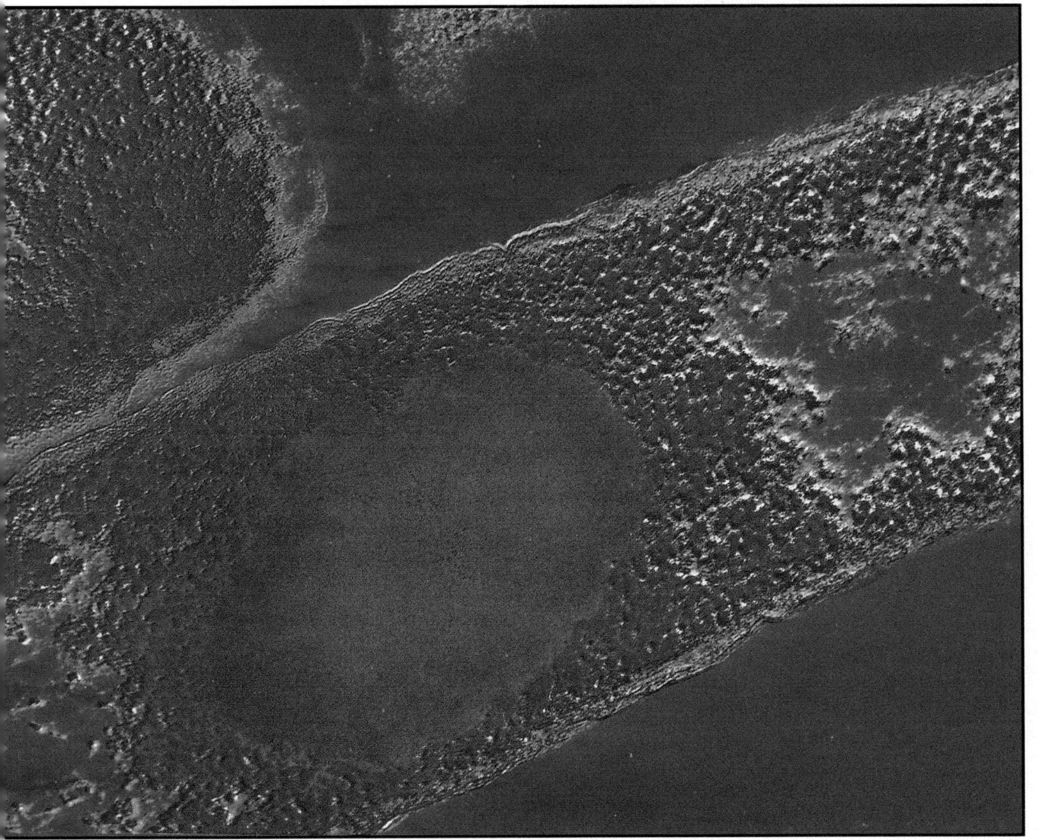

The reddish spot at left is a pool of the protein human gamma interferon inside the common intestinal bacterium *E. coli,* which produced it. To make *E. coli* synthesize the protein, which enhances the immune system, geneticists insert the gene for interferon into the bacterium's DNA, then cultivate the organism in huge numbers to make the protein in quantity.

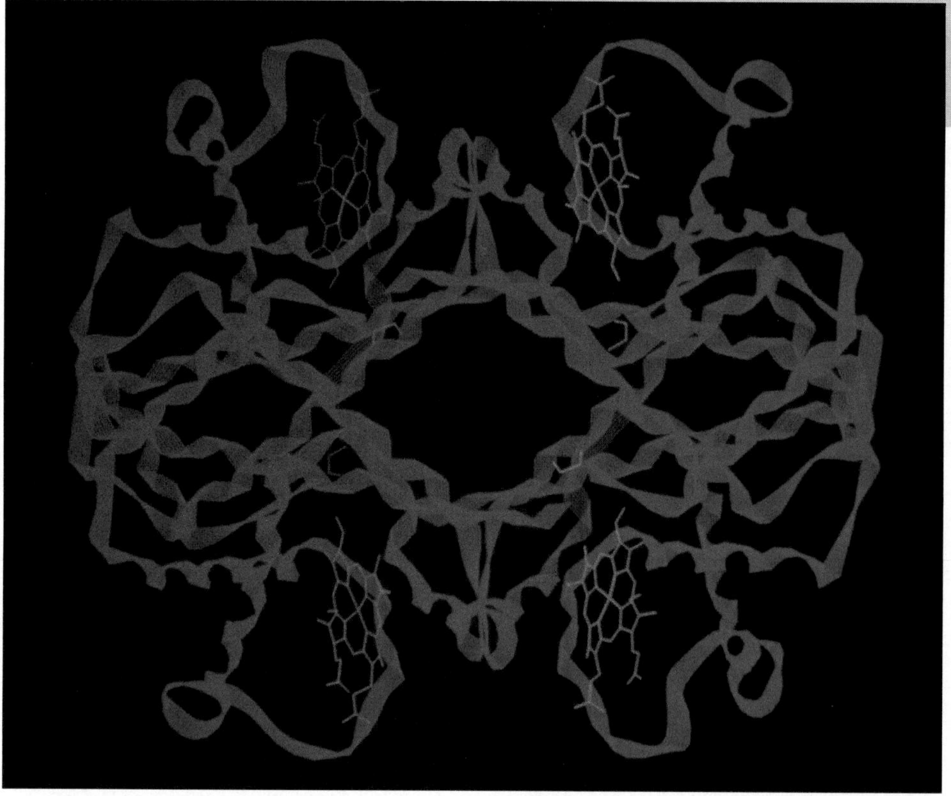

Twisting, ribbonlike strands of blue in this computer-generated model of the oxygen-carrying protein hemoglobin represent two pairs of interlocking protein chains called alpha-globin and beta-globin. The latticelike pink structures depict heme, an iron-rich pigment that binds with oxygen, which the hemoglobin carries from the lungs to tissues throughout the body.

From this position, the provirus normally directs the synthesis of more retroviruses, which then break out of the host cell to infect others. But in Mulligan's plan for a genetically altered retrovirus, a therapeutic gene—such as the one for hemoglobin—would replace part of the harmful provirus. Thus changed, the retrovirus would still be able to penetrate a stem cell and deposit its genetic payload; however, it would not be able to order production of new retroviral particles. Not only would the affected cell begin to produce hemoglobin, but because the virally implanted gene was now part of the cell's own DNA, the gene would be reproduced each time the cell replicated itself. In this way, Mulligan hoped to ensure the efficient spread of his genetic transplant.

Once the scientist had settled on his retroviral truck, he turned to the problem of the genetic freight. A gene isolated for delivery by this method must include DNA control sequences that specify how much of the protein is to be made and which cells are to produce it. In his monkey-virus experiments, Mulligan relied on control sequences within the monkey-virus genome to foster protein synthesis, but he would have to find a different approach before any genetic treatment could work. Without the beta-globin gene's own strips of control nucleotides, even a virus capable of infecting humans would likely cause the wrong cells to produce too little hemoglobin. Worse, the absence of these important elements could conceivably cause white cells to produce hemoglobin, an unnatural state of affairs with unpredictable consequences for the body's immune system.

One of these all-important control sequences is relatively easy to find; it almost always sits right next to the segment of the gene that holds the recipe for its protein. The others, however, can lie a long and variable distance away on the gene.

Molecular biologist Frank Grosveld at the National Institute for Medical Research in England set out in the mid-1980s to make sure that none of the control sequences would be left out of the cargo to be carried by the retroviral truck. To do so, he transplanted human beta-globin genes into mice. With each trial, Grosveld snipped away gene fragments from both ends until he found the shortest length of DNA that also produced copious amounts of beta-globin.

But when Mulligan then attempted to fit this customized cargo into his viral truck, he discovered that the load was too big. Luckily, the recipe section of a gene has many long segments of base pairs that are actually ignored when a protein is manufactured. By getting rid of these introns, as the segments are called—an event that occurs naturally before protein production begins—the beta-globin gene was in the end trimmed to a size that was easily carried by Mulligan's retrovirus.

With his retroviral vehicle and its payload now matched, Mulligan next turned to the painstaking task of testing the gene-delivery system in the lab. Successful experiments on mice in 1987 showed that only about one-tenth of one percent of the human hemoglobin genes inserted into the rodent's genome by the altered retrovirus took root in stem cells and actually began to produce the protein. This was far short of a medically useful result but a breakthrough nonetheless. Never before had a gene been placed in a stem cell: stem cells are few in number compared with most other cells and relatively inaccessible in the bone marrow.

In time, Mulligan hoped to improve this ratio enough for genetic treatment of anemia to be effective, but meanwhile he was less sanguine about another problem. The vast majority of any transplanted genes plant themselves arbitrarily in the recipient's genome—a potentially disastrous development. If a gene happens to take root in the middle of a structural gene, for example, it could halt the gene's production of its protein,

Red blood cells from an individual with sickle cell anemia *(right)* are angular and rigid instead of round and flexible, making passage through narrow capillaries difficult or impossible. This painful and sometimes fatal form of anemia results from a genetic fault that causes hemoglobin molecules to bind together, leading to the misshapen corpuscles.

with results as dire as those of the disease under treatment. Worse yet, a damaged regulatory gene might unleash the production of unwanted proteins or even tip the cell toward chaos and cancer.

As Mulligan labored to find a safe and effective treatment for sickle cell anemia, scientists elsewhere sought to apply his method to other genetic diseases. The earliest successes would come from three doctors at the National Institutes of Health (NIH) in Bethesda, Maryland: W. French Anderson of the National Heart, Lung, and Blood Institute and R. Michael Blaese and Kenneth Culver of the National Cancer Institute. The doctors had been investigating how they might treat a deficiency of the enzyme adenosine deaminase (ADA). Without ADA, wastes from the body's ordinary chemical processes build up in white blood cells, crippling the immune system of which they are a vital component and leaving victims vulnerable to a slew of life-threatening diseases.

Anderson first drew up a treatment plan in 1987. During the next three and a half years, he logged 90-hour work weeks shepherding the proposal through the maze of federal agencies and scientific advisory boards that were assigned to evaluate the risks. Meanwhile, he and his colleagues also performed a remarkable experiment in the laboratory. Using Mulligan's gene-transfer technique, they placed healthy ADA genes into white blood cells taken from four-year-old Ashanthi DeSilva and nine-year-old Cynthia Cutshall, two little girls from Ohio who were suffering from ADA deficiency. To the researchers' joy, the treated white cells duplicated earlier successes with similar cells from animals and began to produce the missing protein in quantities great enough to do some good.

Anderson and his colleagues fully understood the unintended consequences that could result from treating ADA deficiency genetically. However, with their proposal under review by the Human Gene-Therapy Subcommittee at NIH, the three researchers argued persuasively that their laboratory results had a good chance of being reproduced in the human body and that the risks of a bad outcome were extremely low. Furthermore, if the trial were to go awry genetically, they pointed out, the affected white blood cells would all die off within several months, leaving the situation no worse than it had been at the start. The time had come, said the researchers, to test some form of genetic therapy, and no other disease

Message-Bearing Genetic Invaders

Viruses are among the most potentially deadly—and potentially useful—carriers of genetic material. Their makeup is extremely simple: a protein coat surrounding a nucleic acid core that contains instructions for replication but little or no machinery for carrying out those instructions.

Viruses thus are parasitic, requiring tools supplied by a host cell to ensure their own survival. The damage they may inflict on the host in the process varies from something as minor as the common cold to fatal illnesses like AIDS. But as scientists have begun to learn more about these biological machines, they have found ways to manipulate them for medical purposes, turning viruses into efficient mechanisms for correcting genetic flaws at the DNA level (*pages* 96-105).

As shown schematically on the following pages, reproductive cycles vary according to the type of virus. Bacteriophages, seen attacking an E. *coli* bacterium at right, contain double-stranded DNA; other DNA-carrying viruses may contain single-stranded DNA. The so-called retroviruses—including HIV, the virus that causes AIDS—carry single-stranded RNA and the enzyme reverse transcriptase, for converting the viral RNA into DNA shortly after infection. Modified retroviruses from mice are often used to convey therapeutic genes into targeted cells in human patients.

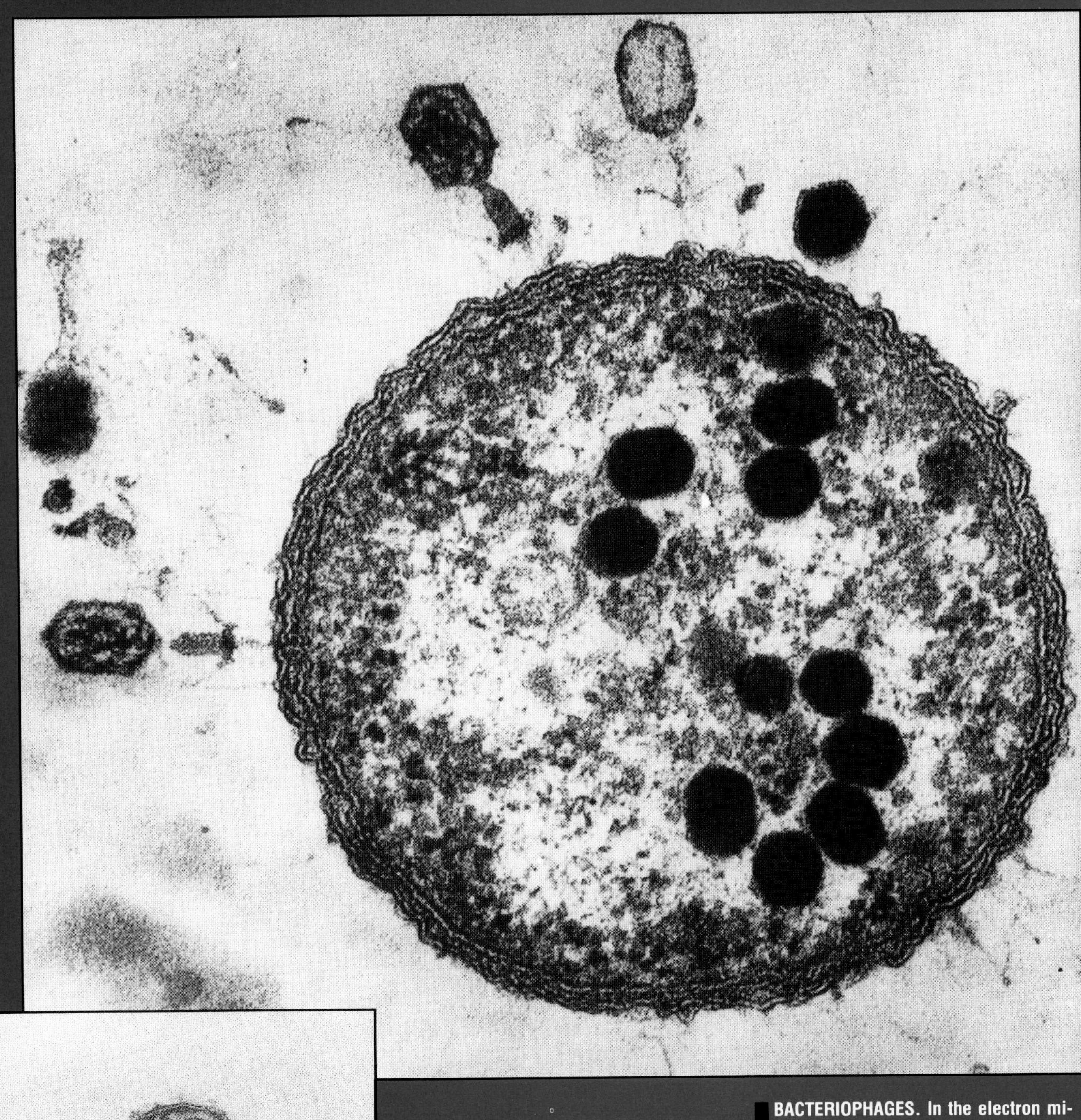

BACTERIOPHAGES. In the electron micrograph above, an army of bacteriophages lay siege to a bacterium. Thirty minutes after the initial infection, newly produced viral particles are visible inside the cell.

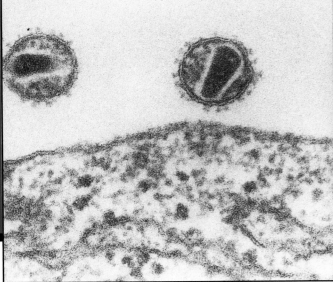

RETROVIRUSES. Encased in a fatty envelope, two AIDS viruses *(left)* approach a T-cell lymphocyte—a white blood cell vital to the body's immune response. Knobby projections called spike proteins bind to receptors on the T cells to initiate the process of insinuating the virus and its RNA into the T cell.

Life Cycle of a Bacteriophage

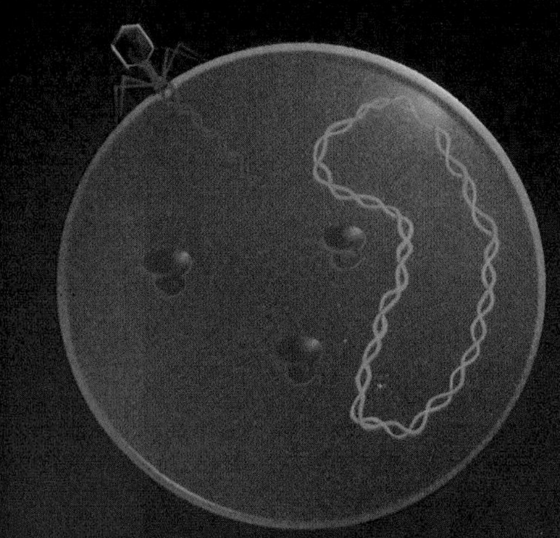

INJECTION. When the bacteriophage comes in contact with a bacterium, appendages called tail fibers hold fast to the cell's surface while the syringe-like protein tail injects the viral DNA, shown in red. Bacterial ribosomes are depicted in purple, bacterial DNA in blue.

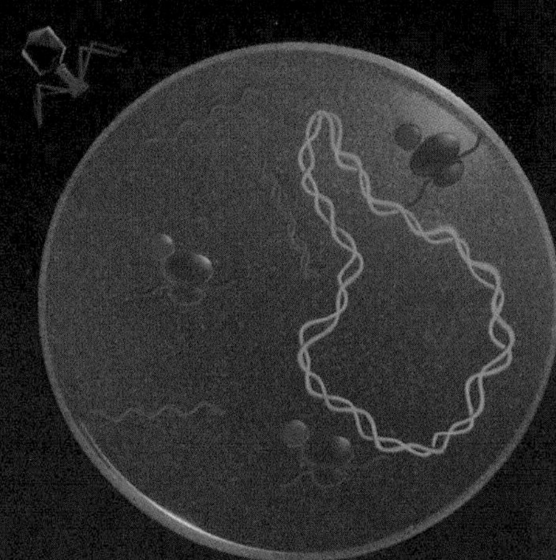

TRANSCRIPTION. Bacterial enzymes transcribe the viral DNA into single strands of messenger RNA. The mRNA then instructs the ribosomes to manufacture necessary viral enzymes *(red spheres)*. Meanwhile, the empty viral shell falls away from the cell surface.

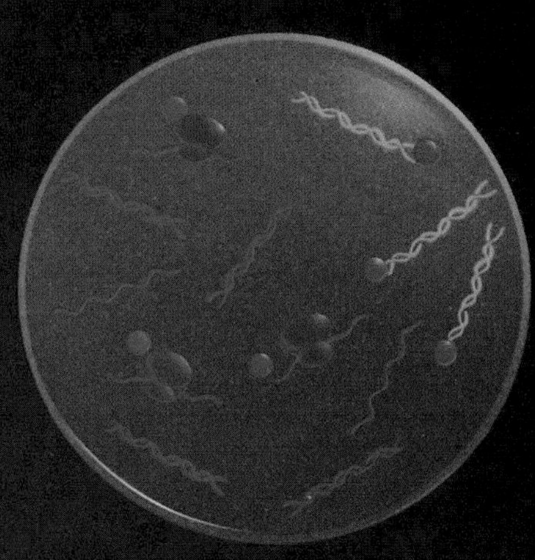

REPLICATION. Of the newly minted viral enzymes, one type destroys the host DNA; another type replicates the viral DNA hundreds of times over. To obtain the necessary nucleotides to form viral DNA, the enzymes cannibalize the remnants of the host DNA.

Life Cycle of a Retrovirus

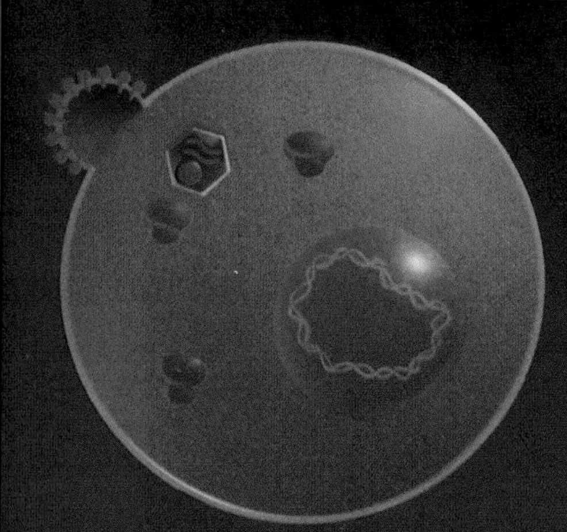

FUSION. Spike proteins in the outer fatty envelope of the retrovirus bind to surface receptors on the host cell, and the viral envelope fuses with the cell membrane. The viral core, or nucleocapsid, carries viral RNA and enzymes such as reverse transcriptase *(red)* into the cell.

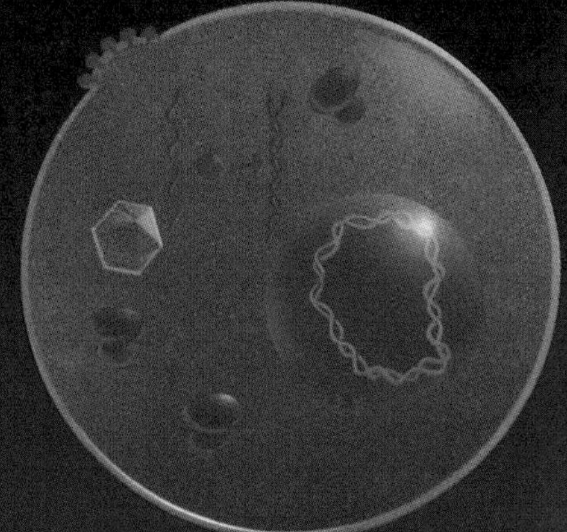

RNA CONVERSION. The cell digests the viral envelope and the protein coat of the nucleocapsid, releasing the viral RNA and reverse transcriptase enzyme. Using the RNA as a template, the enzyme manufactures a complementary double-stranded viral DNA segment *(red)*.

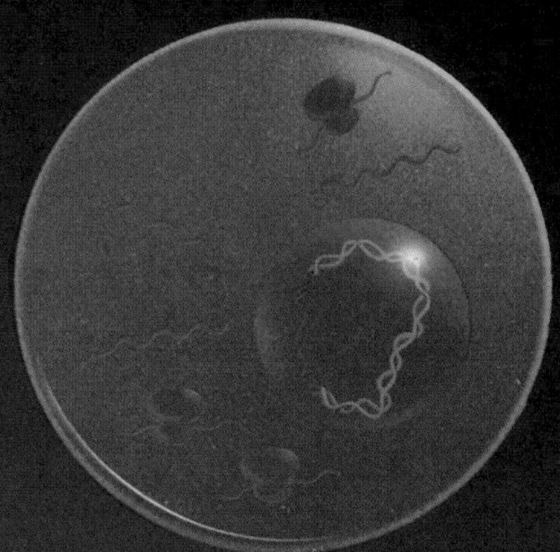

TAKEOVER. The viral DNA *(red)* is integrated into the cell's DNA *(blue)*. As the cell transcribes this combination DNA, called the provirus, it will produce strands of viral RNA, some of which will serve as viral messenger RNA that will be translated by the cell's ribosomes.

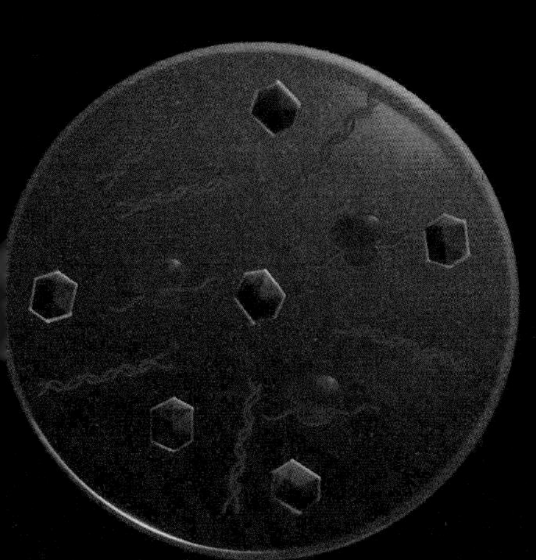

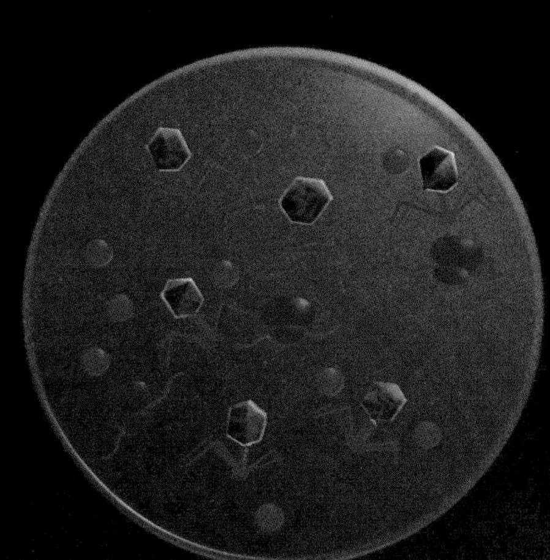

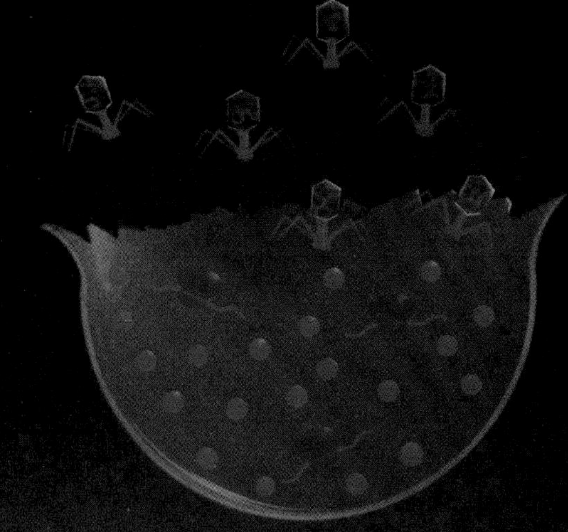

YNTHESIS. The ribosomes continue to follow e messenger RNA's instructions, turning out ral proteins that will form the protective oats of new viruses.

ASSEMBLY. Coat proteins encase DNA and assemble with tails and tail fibers into new viruses. Ribosomes then produce an enzyme called lysozyme, which causes the host cell to break down.

RELEASE. About 30 minutes after infection, the bacterium ruptures from the effects of the lysozyme. New viruses erupting from the host cell are now free to infect other bacterial cells in the vicinity.

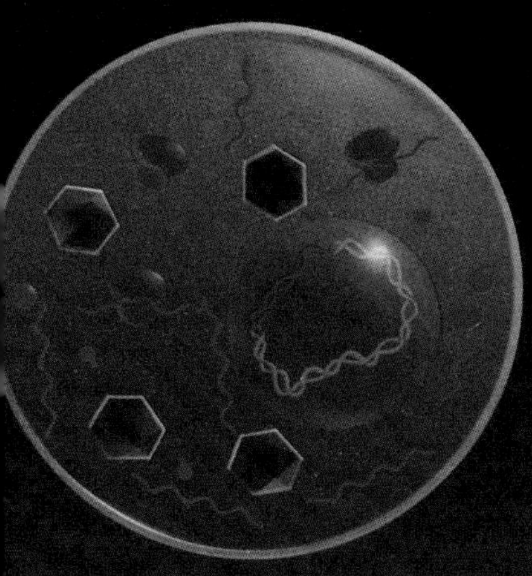

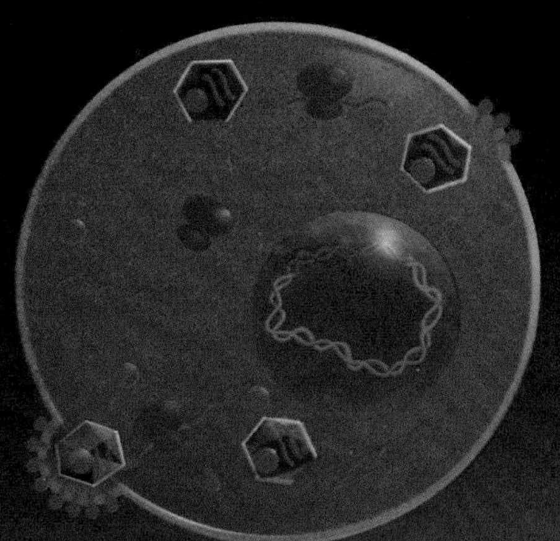

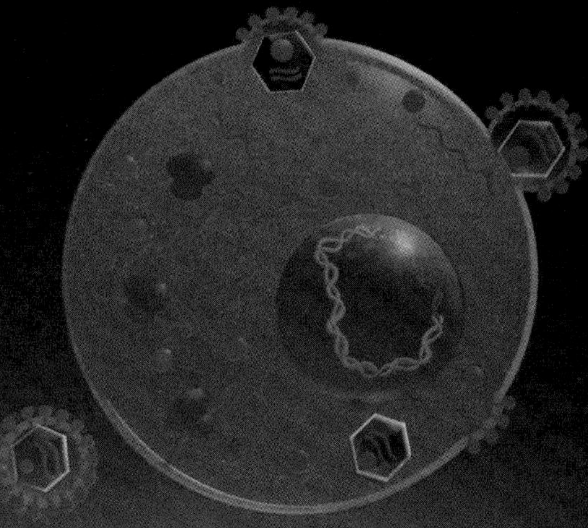

YNTHESIS. Working from the viral messenger NA, the host cell's ribosomes produce sever- l enzymes, including reverse transcriptase, s well as a number of viral structural pro- eins, including the spike proteins of the enve- ope and proteins for the nucleocapsid.

ASSEMBLY. Enzymes and strands of untranslated viral RNA come together inside a protein shell to form viral nucleocapsids, while the spike proteins migrate to the cell membrane, where they will eventually be incorporated into the viral envelope.

BUDDING. The nucleocapsids bud from areas on the cell membrane studded with spike proteins. As a nucleocapsid is released, it is wrapped in a fatty envelope consisting of the cell membrane and spike proteins. New retroviruses continue to bud for the life of the cell.

offered better prospects of success or a greater benefit if it worked.

Confident that the committee would give its approval, Anderson and his colleagues removed a billion or so white blood cells from Ashanthi DeSilva and exposed them to tamed leukemia retroviruses carrying the human ADA gene. As expected, the viruses invaded the cells and obediently stitched the foster gene into the cells' chromosomes. The researchers then mixed the cells into a sterile saline solution for storage pending final review of their proposal.

In the end, the Human Gene-Therapy Subcommittee did give their consent, and the researchers were allowed to go forward with a human test. Only Richard Mulligan, a member of the committee, disapproved. Clinical trials on humans should wait, counseled Mulligan, until more of the consequences of such treatments were understood. "I don't want to be a piece of history," he said at the time, "that's moved too fast, that's rushed to do things in humans, that's been careless. We want to have a clear idea of the risks and benefits before we give a gene transplant to a patient. First we should do careful statistical studies with animals—studies that have yet to be done—and do everything we can to check a virus's safety." But Mulligan went further. Calling the proposal "technically and scientifically a bad idea," he lambasted it for therapeutic shortsightedness. Because the genes were to be implanted in mature white blood cells, which have a life expectancy of no more than several months or so, the therapy would have to be repeated at regular intervals.

Mulligan advocated postponing the experiment until a technique could be perfected for isolating stem cells. Placing the ADA genes there would be a one-time affair, assuring that every white cell coming from a genetically repaired stem cell would have the functioning ADA gene. But instead of waiting until a method for collecting stem cells could be found, Anderson and his fellow researchers opted "to start with a target cell we knew we could hit."

On September 14, 1990, the doctors transfused the white-cell mixture into Ashanthi's hand in a procedure that took about half an hour. Four-and-a-half months later, after it had become clear that Ashanthi's health was improving without significant side effects, Cynthia Cutshall received a similar treatment.

Early assessments of the therapy's efficacy indicated that Anderson and his coinvestigators were right to proceed. More than a year after the treatment, both patients were enjoying robust good health. One child even resisted an outbreak of chickenpox that swept her school. "We all held our breath," says Blaese. "It was an acid test of the treatment."

By 1992, a new drug was discovered that permits gene therapists to flush stem cells out of their usual haunt in the bone marrow and into the blood, from which the cells can be isolated and, it is hoped, their genes repaired. In early May 1993 Cynthia Cutshall became the second person to undergo an attempt at supplying her stem cells with functioning ADA genes and then returning the cells to her body. (The first was an Italian child, whose name was not made public.)

All the participants crossed their fingers that the stem cells would take up the new gene and migrate back to the bone marrow, there to produce white blood cells capable of manufacturing enough ADA to solve the children's immune-system problems once and for all. In a sense, this experiment was riskier than the earlier effort with white blood cells, because if the genes entering the stem-cell DNA had an undesired side effect, the damage would be built into all the white cells spawned for years to come by the affected stem cells.

Nonetheless, Anderson, Blaese, and Culver had opened important doors.

The Vagaries of Jumping Genes

At a press conference following the announcement of her Nobel Prize, Barbara McClintock shows off an ear of the corn that cemented her reputation as a geneticist. Ignored for decades, she persevered in her work all the same. "When you know you're right," she said, "you don't care what others think. You know sooner or later it will come out in the wash."

A rare mutation from too much sun or some other external cause notwithstanding, it would seem that DNA replication and gene expression (*pages* 39-49) are models of fail-safe processes, dependably assembling identical copies of DNA and of the proteins for which DNA holds the formulas. The truth, however, is not so neat.

Genes were once thought to occupy only a single position on a DNA strand, but now geneticists know that some genes can change positions during replication and perhaps at other times. Not all the effects of these mutations are known, but one of them is to prevent an adjacent gene from being expressed. Hemophilia, for example, can be caused by a gene that moved from its position on the parents' DNA to take up residence in the child next to a vital blood-clotting gene, turning it off.

Jumping genes like this one were first proposed by geneticist Barbara McClintock (*left*) in the 1940s. While conducting breeding experiments with Indian corn, or maize, she noticed that the colors of some kernels deviated inexplicably from the rules of genetics established by Gregor Mendel.

McClintock correctly attributed the aberrations she saw to migrating genes, but the idea was ridiculed as heresy by her colleagues until the 1970s, when molecular biologists saw some of the wayward genes she had predicted. Vindicated at last, McClintock won the 1983 Nobel Prize in biology for her insight, ranked by the prize committee as second in importance only to the discovery of DNA's double-helix structure.

Following their trailblazing work in treating ADA deficiency, the NIH approved human trials of some 50 genetic therapies. Among them was one that targets cystic fibrosis (CF), the most widespread genetic disease among people of European ancestry. Victims of CF lack a protein that scientists believe helps to regulate salt balance in the lungs. Without this protein, which has a tongue twister of a name usually shortened to CFTR, salt accumulates in lung cells. There it absorbs water from the mucus coating inside the lungs, making the coating thick and sticky. The altered consistency impairs breathing and promotes infection. Most CF sufferers die before the age of 30.

In 1993, two dozen or so American CF patients were treated with a solution containing disabled cold viruses that had been equipped with healthy CFTR genes in the laboratory. As the solution was dripped into the patients' lungs, the virus burrowed into lung cells, where its adopted gene began producing the missing CFTR protein. Scientists hope that the level of protein synthesis will be sufficient to reverse the disease. "There is a great deal of testing that has to be done to evaluate the safety and efficacy of the therapy," cautioned NIH researcher Ronald Crystal, who devised the therapy one spring day while out for a run. "I cannot guarantee that this is going to be the cure for the disease."

For the foreseeable future, scientists expect that gene doctoring will be limited to disorders—such as ADA deficiency and cystic fibrosis—that are caused by a single defective gene. Most genetic disorders, however, result from the misbehavior of several genes. Arteriosclerosis and diabetes are common examples, but probably the best known and most feared of these afflictions is cancer. Until recently, it was also one of the least understood.

The genetic underpinnings of human cancer were not exposed until around 1980, when lab experiments showed that DNA taken from a tumor could convert normal cells into cancerous ones. Careful analysis of the tumor DNA revealed that the cancer-inducing portion was confined to a tiny segment—what appeared to be a single gene.

Investigators tailored a probe to seek out the gene's position in the healthy human genome. The probe bound to a stretch of DNA that differed from the tumor probe in only one of approximately 5,000 base pairs. Clearly, a single-point mutation within a normal gene had transformed it into a killer called an oncogene. Yet this discovery told scientists nothing of why the normal gene had mutated—or how that mutation had driven the cell toward the unbridled growth known as cancer.

In 1982 Bert Vogelstein, a young molecular biologist working at the Johns Hopkins Oncology Center in Baltimore, Maryland, set out to solve this riddle for colon cancer. Before he could hope to unravel the secrets of the disease, however, Vogelstein first had to identify the mutated genes that bring it on.

The problem was that Vogelstein had no idea where to start looking. Lacking a better approach, he began by testing DNA from human tumors of the colon with oncogene probes fashioned from cancers in animals, hoping that these DNA tracers would match human oncogenes closely enough to attach to them. When this strategy yielded nothing, Vogelstein tried to make the aberrant genes reveal themselves: He extracted DNA from human colon-cancer tumors and mixed it with healthy cells in a petri dish. That the cells remained normal was a disappointment.

Vogelstein began to wonder if colon

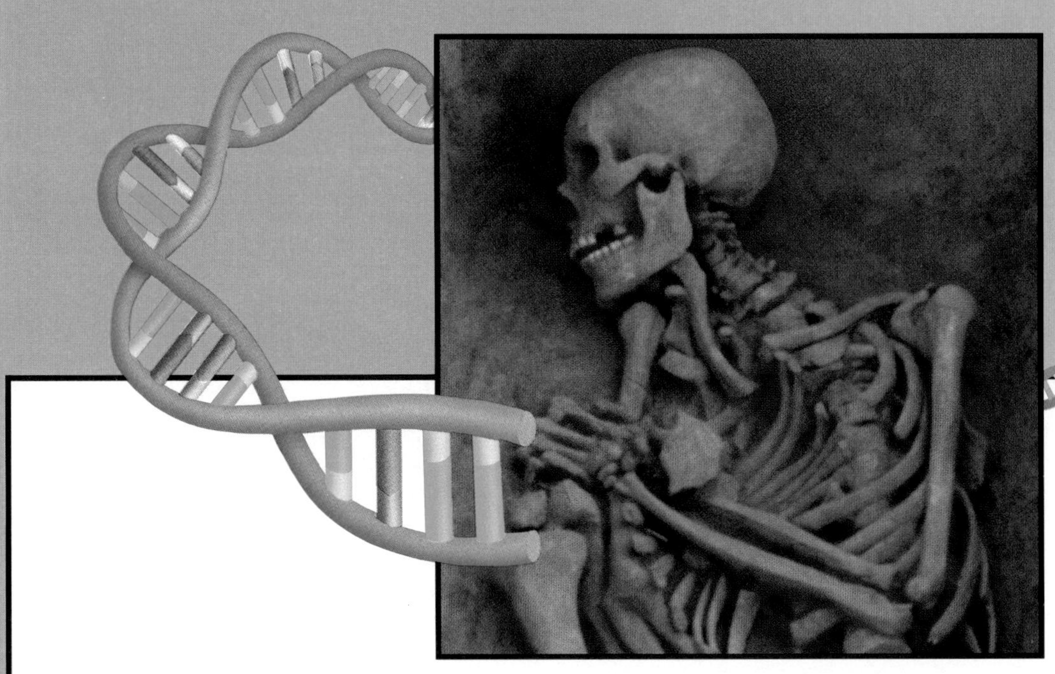

Positive Identification through a DNA Profile

That a human pedigree, unsupported by any written records whatever, could be traced 400 generations into the past seems improbable at best. Yet genetic technicians have done exactly that in the case of Native Americans in North and South America. A close analysis of samples of DNA from living Indians, from prehistoric Indian skeletons like the one above—recently exhumed in the United States—and from Siberians living across the Bering Strait from Alaska suggests that some of the early inhabitants of the Americas traveled from Asia to Alaska some 12,000 years ago, perhaps across a land bridge exposed during the last ice age.

The technique behind this discovery is called DNA profiling. Developed in 1985 by Alec Jeffreys at the University of Leicester in England, the process reveals minuscule details of an individual's genetic makeup and was originally intended to detect the presence of genes for inherited diseases. But its utility has spread far beyond medical applications, and DNA profiling—sometimes called DNA fingerprinting or typing—is now used to nab criminals, to settle paternity disputes, to provide positive identification of fallen soldiers, and in the field of anthropology to understand the origins of peoples and their routes of migration.

To trace the lineage of Native Americans, researchers used mitochondrial DNA (*page* 64), which is transferred from generation to generation through the mother, unaltered by genes from the father. Exhaustive analysis provides evidence that all full-blooded Indians in the Americas are descended from just four ancient maternal lineages. Identifying individuals, however, requires DNA from cell nuclei, a double helix combining contributions from both father and mother. The method most often used to identify criminals is described on the following pages. It relies on the fact that some sequences of nucleotides in genes are repeated many times, forming chains of repetitions that differ in length from one individual to the next.

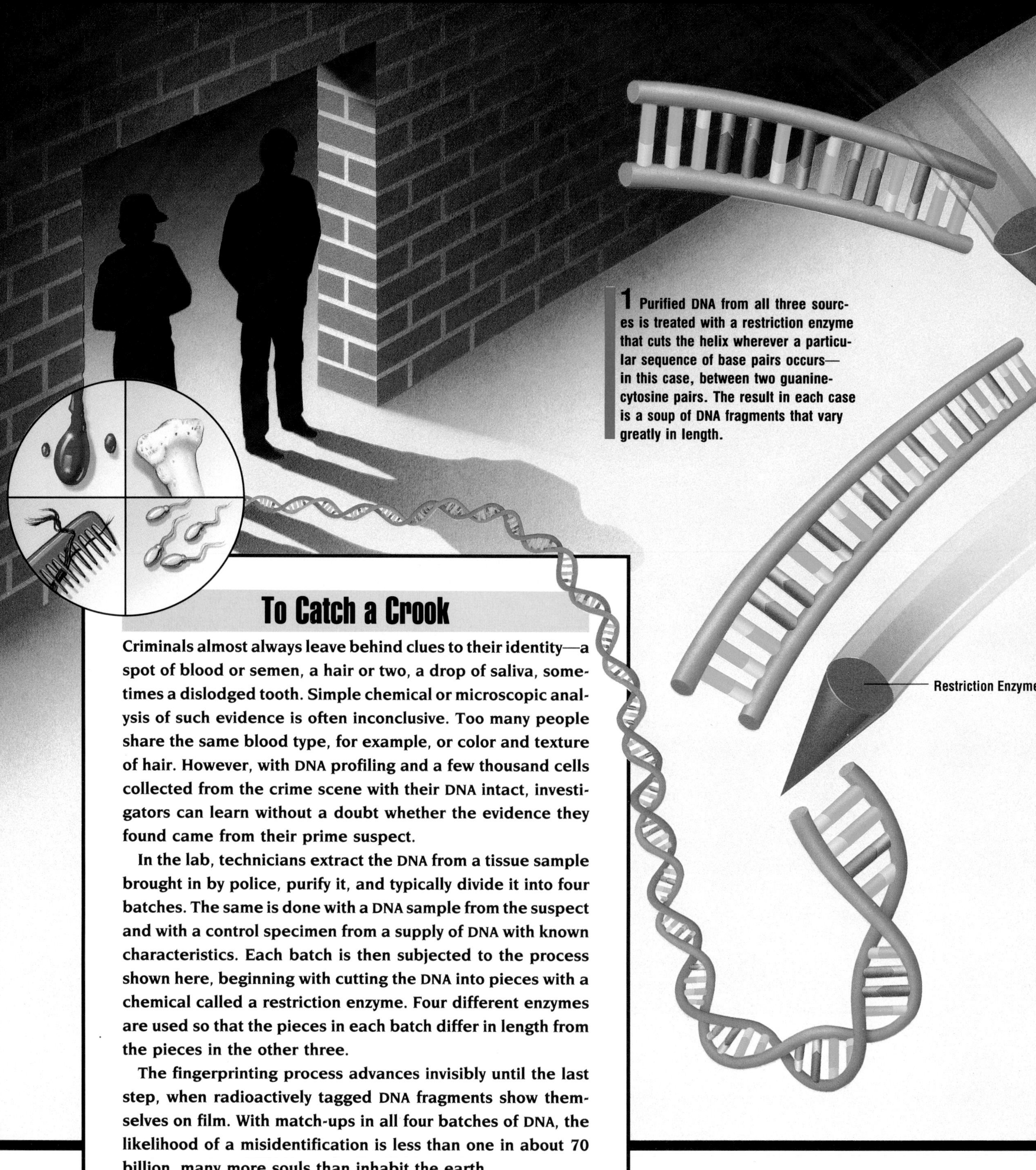

1 Purified DNA from all three sources is treated with a restriction enzyme that cuts the helix wherever a particular sequence of base pairs occurs—in this case, between two guanine-cytosine pairs. The result in each case is a soup of DNA fragments that vary greatly in length.

To Catch a Crook

Criminals almost always leave behind clues to their identity—a spot of blood or semen, a hair or two, a drop of saliva, sometimes a dislodged tooth. Simple chemical or microscopic analysis of such evidence is often inconclusive. Too many people share the same blood type, for example, or color and texture of hair. However, with DNA profiling and a few thousand cells collected from the crime scene with their DNA intact, investigators can learn without a doubt whether the evidence they found came from their prime suspect.

In the lab, technicians extract the DNA from a tissue sample brought in by police, purify it, and typically divide it into four batches. The same is done with a DNA sample from the suspect and with a control specimen from a supply of DNA with known characteristics. Each batch is then subjected to the process shown here, beginning with cutting the DNA into pieces with a chemical called a restriction enzyme. Four different enzymes are used so that the pieces in each batch differ in length from the pieces in the other three.

The fingerprinting process advances invisibly until the last step, when radioactively tagged DNA fragments show themselves on film. With match-ups in all four batches of DNA, the likelihood of a misidentification is less than one in about 70 billion, many more souls than inhabit the earth.

2 In a process called gel electrophoresis, DNA fragments in each sample form a track on a bed of gelatin, arranging themselves by length in response to an electric current. An alkaline solution is then applied to split the double-stranded DNA fragments into single strands *(inset).*

3 In a technique called Southern blotting, a thin film of nylon is laid atop the gel to blot up the DNA fragments and fix their positions relative to one another. Care must be taken not to stretch the nylon; such distortion can lead to invalid results.

4 The nylon is immersed in a bath containing many copies of a radioactive probe, a sequence of nucleotides that binds with its complementary sequence wherever it occurs in the DNA *(inset).* The resulting clusters of radioactivity fog a sheet of x-ray film placed against the nylon.

5 When developed, the film shows two smudges for each sample. Positions of smudges from the suspect's DNA and DNA collected at the crime scene clearly indicate a match not even approximated by smudges from the control sample. Tone differences in smudges are insignificant.

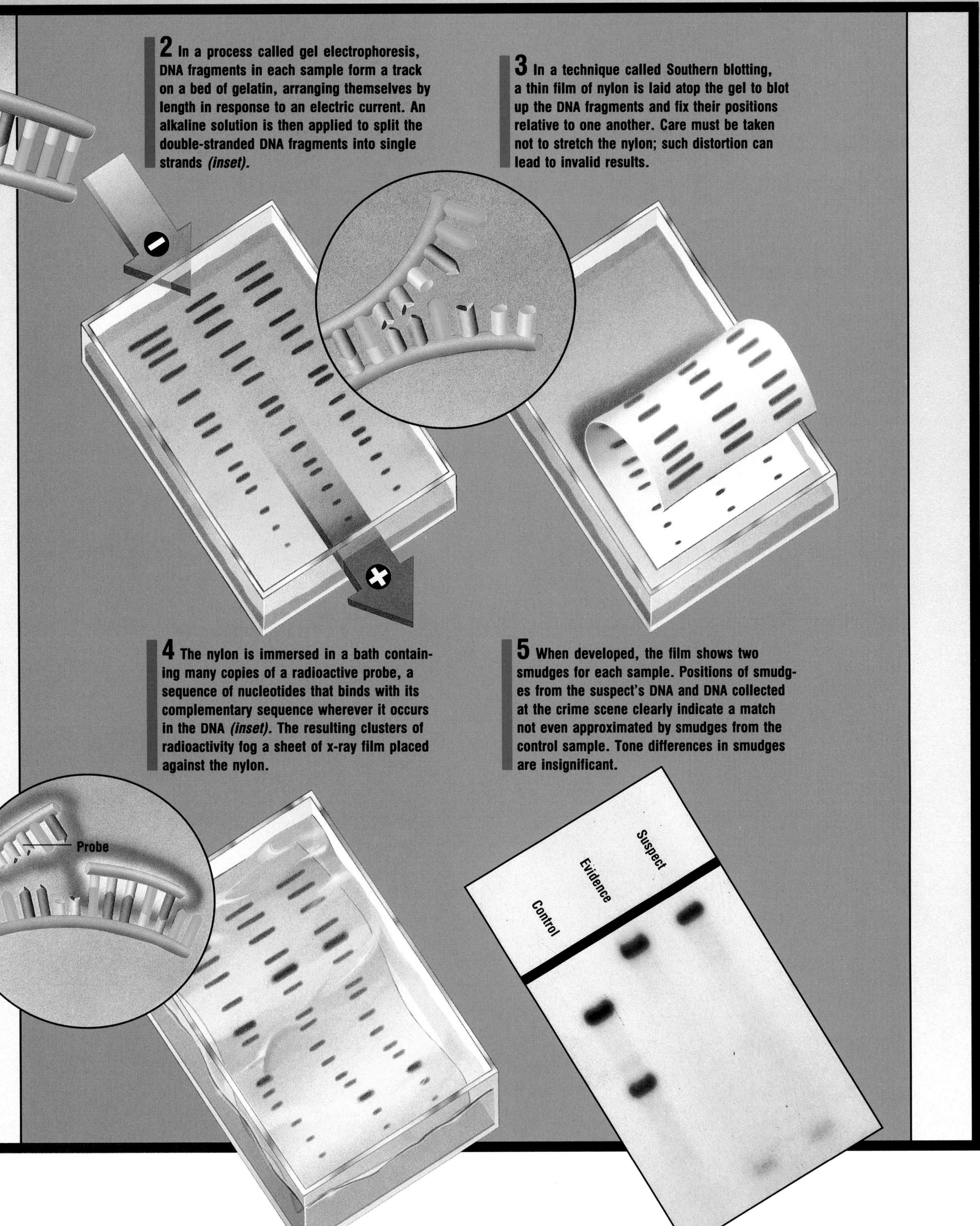

cancer had a genetic basis after all. Then it occurred to him that some of the DNA he was using might have come not from cancerous cells but from other, healthy cells in the immune system or from undiseased connective tissue in the colon.

So in 1985, Vogelstein and a team of researchers assigned themselves the tedious task of coaxing apart colon tumors cell by cell, separating cancerous from noncancerous tissue. Nearly a year and hundreds of tumors later, they had enough unadulterated tumor DNA to resume research. This time, instead of using probes from animal tumors, the Hopkins investigators unleashed radioactively tagged probes for human oncogenes recently isolated from colon tumor cells. Like a homing pigeon, one probe made straight for a gene sequence on chromosome 12. It was a mutated ras gene, a class of oncogene similar to one identified in human bladder tumors a few years earlier.

Research on ras oncogenes had revealed that these mutants contribute to cancer in a number of ways that can easily foment unchecked cell division. But not all of the colon tumors Vogelstein analyzed carried the ras oncogene. Young emerging cancers

This image of a lung-cancer cell, its DNA catastrophically altered, was made using a scanning electron microscope, then tinted by a blue filter. With bloated, lumpy contours, the diseased cell is readily distinguishable from a healthy one, which has a flat, scaly appearance.

bore no trace of it. When probes for other oncogenes turned up no matching sequences in the tumor DNA, Vogelstein had to concede that oncogenes were only part of the story. Perhaps, he thought, damage to—or elimination of—another type of gene accounted for the disease's earlier stages. To test his theory, Vogelstein and his associates began scrutinizing chromosomes of tumors at various stages of development for jumbled or missing DNA sequences. By 1987 they had located three: absent segments on chromosomes 5 and 18, and a mutation on chromosome 17.

What most fascinated Vogelstein was that each of the chromosomal anomalies showed up in the tumor DNA at a different point in the disease cycle. In an effort to understand how each muddled gene contributed to the cancer's progress, the team isolated the responsible genes, starting with chromosome 17—site of the single most frequently observed mutation. By checking flawed copies of the chromosome against normal samples, they zeroed in on a gene with one erroneous base pair.

It was p53, guardian of the genome against the depredations of ultraviolet radiation from the sun and of other agents that provoke cancer-causing mutations. A single-point mutation, it seemed, had destroyed the gene's ability to arrest abnormal cell development. Vogelstein was dubious. He decided to see what happened when he placed normal p53 genes in cancer cells growing in the laboratory. The results were as gratifying as they were unexpected. Said Vogelstein: "The gene is only one of several that are mutated in colon cancer. So you could imagine that if you put back a normal copy, it might have a slight effect. But it had a huge effect. It completely stopped tumor growth."

The other genes on chromosomes 5 and 18 turned out to be "tumor suppressor" genes as well. Gradually, by noting at which stage of the disease a particular mutation appeared, Vogelstein was able to piece together a genetic chronology for colon cancer.

In the earliest phase of the disease, the tumor suppressor gene on chromosome 5 is inactivated, permitting a tiny, benign growth called a polyp to develop from cells lining the bowel. Next, a ras oncogene accelerates the abnormal growth of the polyp. In phase three, a tumor suppressor gene on chromosome 18 is knocked out and the polyp balloons in size. Finally, in stage four, p53 sustains damage. The unregulated tumor then turns malignant: Instead of being confined within itself, the growth begins to spread, penetrating the colon wall and invading other parts of the body.

Follow-up studies indicated that mutations of the p53 gene precede the final assault by a variety of other cancers, including those of the lung, breast, ovaries, cervix, brain, bone, and bladder. "It turns out," Vogelstein reported, "that p53 is currently the most frequently mutated gene known to exist in cancer."

Even so, many cancers appear to result from a series of genetic mishaps, no one of which causes the disease by itself. Scientists have identified three ways in which a normal gene can become precancerous. Often, environmental hazards such as radiation or chemical carcinogens cause a gene to mutate. If a gene switches positions on a chromosome or jumps spontaneously from one chromosome to another (*page* 75), it can begin to overproduce its protein, disrupting cellular mechanisms in the bargain. Finally, a jumping gene or virus may trigger the runaway production of protein by the gene that it lands next to.

To complicate matters even further, Vogelstein and scientists at the University of Helsinki in Finland have discovered a mutant gene that seems to actively promote colon cancer and other forms of cancer as well. It is not the result of any environmental carcinogen; rather, this mutation is an inherited one. From early in life, the aberrant gene appears to go on a rampage, actively causing countless other mutations that eventually overwhelm tumor suppressor genes such as p53. Cancer is the result.

Findings such as these offer insights about cancer, but they are narrow ones. Until a great deal more is known about how all these factors and others cooperate to foster cancer, a gene-therapy cure for these tragic diseases will remain out of reach. Still, as Vogelstein points out, "The door has opened a crack." It is only a matter of time.

The wait would be considerably shorter if every gene's position on its chromosome, as well as its function and its base-pair sequence, were known. For instance, much laboratory time now devoted to sending probes to seek out genes could be saved. Instead, a powerful computer could review the entire genome for a sought-after sequence of nucleotides.

Bits and pieces of this huge and complex puzzle have emerged over the years, but until recently there was no coordinated effort to speed progress. Historically, biologists the world over have worked in isolation. Their science is a cottage industry—a loose association of independent labs in which a couple of postdoctoral students and a senior scientist toil away on a shoestring grant to solve one of nature's biological riddles.

By necessity, such labs have practiced a piecemeal approach to deciphering the human genome—targeting one gene here, another there. Moreover, researchers have used different techniques to find and catalog the genes they study, making information sharing a difficult and sometimes fruitless exercise.

But no more. In 1990 the world's genetic-research community banded together in an unprecedented bid to streamline genome research. A U.S.-sponsored international confederation of scientists and genetic-research labs is systematically tackling the job of decoding the entire three-billion-letter text of the human genome. Known as the Human Genome Organization (HUGO), the alliance has set up as its ultimate goal the base-by-base sequencing of all human chromosomes.

More immediately, HUGO's aim has been to produce a linkage map of the genome showing, among other things, approximately where on the chromosomes genes of genetically transmitted diseases reside. In creating a linkage map, researchers use molecular knives called restriction enzymes to cut the DNA of chromosomes into fragments. Since a given restriction enzyme consistently cuts the DNA double helix between two particular bases, the lengths of the pieces of a mutation-free chromosome are different from those of a mutated, disease-carrying chromosome with a dissimilar sequence of bases. Knowing this, genetic mapmakers can plot the location of defective genes to within about 10 million bases. The precision of a linkage map, although far less than the genome project is eventually aiming for, is adequate in most instances to determine whether a child has inherited a gene that caused disease in a parent.

The first and most complete linkage map to come out of HUGO was generated by a Paris-based laboratory established in the mid-1980s, after its director, Tunisian-born physician Daniel Cohen, met with Bernard Barataud, president of France's muscular dystrophy association (AFM). Dismayed at the slow pace of the search for muscular dystrophy genes, Barataud prompted Cohen to collaborate with a French engineering firm to mechanize what Cohen refers to as "the monkey work" of DNA processing. With hefty financial backing from AFM, Cohen coordinated the tooling and start-up of the new automated center for genome research. It was named Généthon, in honor of AFM's

Checking the Genes of Unborn Babes

Since the 1960s, several prenatal tests have been developed to detect abnormalities in fetuses. Most tests require a complete set of chromosomes from a fetal cell acquired by one of the procedures described below. Technicians can count the chromosomes to be sure all 23 pairs are present, and they can tell almost at a glance whether the child will be a boy by the presence of the distinctively short Y chromosome. Closer examination of the chromosomes reveals genetic disorders such as cystic fibrosis, muscular dystrophy, and Down syndrome.

AMNIOCENTESIS. Fluid drawn from the amniotic sac between the 15th and 18th weeks of pregnancy contains fetal cells cast off by the growing baby. In addition to performing genetic tests, technicians can check the concentration of alpha-fetoprotein (AFP). An abnormally high level of AFP can signal a spinal cord that is not fully enclosed by the spine. Low AFP often indicates the presence of Down syndrome.

CHORIONIC VILLUS SAMPLING (CVS). In this procedure, a small sample of tissue is taken from the fetal side of the placenta. Although somewhat riskier to the fetus than amniocentesis, CVS can be done up to five weeks earlier.

BLASTOMERE ANALYSIS. In this experimental procedure, an egg fertilized in the lab for later implantation in the womb—a test-tube baby—can be checked genetically at the blastomere stage, when the embryo has only eight cells. Though one cell is sacrificed for the test, the fetus develops normally.

CELL SORTING. Projected for the late 1990s, this test poses no risk to the child. Blood from the mother, which always carries some cells shed by the fetus, is passed through a high-speed cell sorter to isolate the fetal cells, which are then analyzed to determine the state of the baby's genes.

annual muscular dystrophy telethons.

In 1992, nearly three years ahead of schedule, the genome center astonished American and European labs by publishing a linkage map of the entire human genome. According to Cohen, Généthon owed its stunning achievement to its industrial approach. On the lab's third floor, occupying nearly an acre of floor space, are 20 cube-shaped robots arrayed around a mainframe computer. In one day—and with little human assistance—the robots sort and identify thousands of DNA fragments, work that would require the efforts of technicians sufficient to staff 60 large, less-automated laboratories. More impressive than the machinery, however, are the leaps in knowledge it makes possible.

In October 1992, Généthon announced that it was well on its way to reaching HUGO's second milestone—a contiguity (or contig) map. Sometimes called a physical map, this type of gene inventory charts the exact distances, in numbers of base pairs, among the particular sequences of nucleotides, or markers, next to restriction-enzyme cuts in the DNA. Coming up with a count of base pairs requires the analysis and piecing together of bits of DNA in the correct order. As explained on pages 92-93, scientists need thousands of copies of each segment of DNA to assemble a contig map. Moreover, the longer the fragments, the less sorting and piecing that has to be done.

Knowing this, Généthon scientists adopted an organism called Mega Yeast, a special DNA replicator created in the laboratory by Dr. Maynard Olson during his research pursuits at Washington University in St. Louis, Missouri, in the late 1980s. Mega Yeast cells contain unusually large chromosomes into which more than one million nucleotide base pairs will fit. These Mega Yeast Artificial Chromosomes (Mega YACs) have allowed Généthon to increase its already impressive data-handling rate fivefold. Before computers and Mega Yeast, the equivalent output would have required the efforts of 200 Ph.D.'s. Now, a handful of technicians seated at work stations annexed to the lab's mainframe catalog can analyze 3,200 Mega YACs a day.

Haste in pursuit of a contig map has resulted, as it does in many other endeavors, in a modicum of waste. When large pieces of human DNA are placed in yeast cells, some parts get lost in the process. Just as bad, some of the DNA included can be additions from another chromosome. The differences—between human DNA as it occurs naturally in the body and how it is replicated in yeast cells—can cause errors and confusion, slowing down the mapping process while the mix-ups are resolved.

American genome researchers, who are using shorter sequences of nucleotides grown in bacteria instead of yeast, are poised to fill in some of the gaps: They are using a greater number of restriction enzymes to cut DNA in more places. The outcome promises to be a superior contig map that tells geneticists, with much greater accuracy than earlier maps, where disease-causing genes can be found. Other results include a contig map of the Y chromosome, which dictates maleness, and of chromosome 21, implicated in the cluster of birth defects known as Down syndrome.

HUGO's holy grail, of course, is a list, in order, of all of the estimated three billion base pairs in the human genome. Under the best of circumstances, work on this most detailed of genetic maps is likely to continue well into the 21st century. In its entirety, this so-called Book of Man would fill 10 volumes of the *Encyclopaedia Britannica*.

The Human Genome Project, as the U.S.-based effort is known, promises to spark a revolution in medicine that will spawn marvelous new disease therapies and vastly enrich scientists' understanding of the biochemistry of life. If prognosticators can be believed (and they sometimes prove to have been overly optimistic), billion-dollar industries specializing in biotechnology and pharmaceuticals will flood the market with genetically engineered drugs and sophisticated genetic tests.

In time, some scientists predict, a total "gene screen" will be available. A computerized machine will have the capability to process a drop of blood and build from it a complete genetic profile. Serving as a medical prognosis, the profile would cover everything from whether a child has inherited a genetic predisposition toward any diseases to the presence of latter-day mutations in adults that might increase the chance of contracting maladies such as cancer and hardening of the arteries. Aided by such technology, doctors may be able to prescribe preventive measures that could help patients ward off a host of afflictions.

If, as some people hope and others fear, genetic screening becomes standard practice, the obvious benefits will be accompanied by profound challenges to society's cherished notions of civil liberty, ethics, and personal choice. Past events have already foreshadowed the kinds of dilemmas that cloud the horizon. For example, in the 1970s, a well-intentioned effort to test Americans for sickle cell anemia led to health-insurance cancellations for individuals who were found to carry a recessive gene for the disease. And more recently, a couple from the United States whose unborn child tested positive for cystic fibrosis were ordered by their health maintenance organization to abort the fetus or risk losing medical coverage. Only when menaced with a lawsuit did the HMO retract the threat.

Sobering precedents like these raise concerns about who will have access to genetic information and how it will be used. In the workplace—where, in the United States alone, hundreds of thousands of people suffer from illnesses such as emphysema and cancer—genetic screening could be used to control health-insurance costs by winnowing out job candidates who might be predisposed to such diseases.

What many people find the most disturbing is the prospect that genetic advances will change concepts of what is physically and mentally desirable in a human being. Will mothers who decline to abort genetically impaired fetuses be branded as socially irresponsible? Still worse, will ambitious parents authorize the reengineering of their reproductive cells in

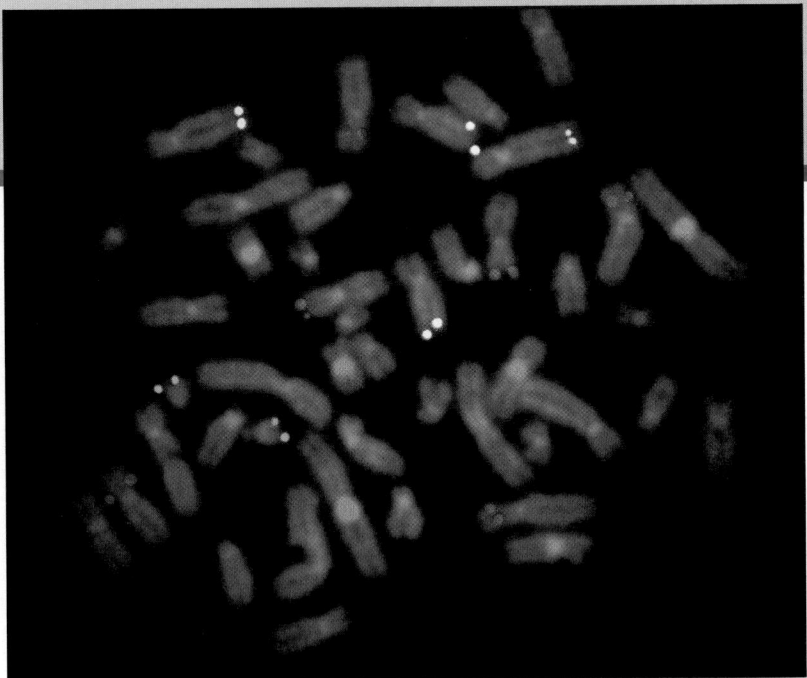

Glowing Beacons for Genetic Testing

The bright-eyed critters in the picture above are human chromosomes harvested just before cell division. Stained blue, they were tagged with half a dozen genetic probes, each carrying a distinctive fluorescent dye. In reality, the dyes emit light in a narrow range of grays, but special filters on a microscope help a computer distinguish between the drab tones, then color them brightly for foolproof recognition.

Fluorescent probes are handy for a variety of purposes, including diagnosis and treatment of cancer as well as prenatal genetic testing. In the chromosomes shown here, two of the six probes warn of possible trouble by tagging the location of specific DNA sequences. The red one marks the donor of the chromosomes as a carrier or potential victim of muscular dystrophy, while the white probe indicates an increased risk for some cancers.

an attempt to produce a brighter, more beautiful child?

"It would be naive to say any of these answers are going to be simple," says James Watson, codiscoverer with Francis Crick of DNA's double-helix structure. "We have to be aware of the really terrible past of eugenics, where incomplete knowledge was used in a very cavalier and rather awful way, both here in the United States and in Germany. We have to reassure people that their own DNA is private and that no one else can get at it. If we don't think about it now, the possibility of our having a free choice will one day suddenly be gone."

This sense of urgency is shared by HUGO's member nations, who have already earmarked 3 percent of the organization's 3.5-billion-dollar budget for the study of the project's ethical implications. Still, among the innovators and visionaries who are driving the genetic revolution, these fears are tempered somewhat by a

Computers *(left)* play a major role in speeding the Human Genome Project toward its goal: a sequence map listing all the nucleotides in human DNA. Part of the map for hemoglobin—about one two-millionth of the entire genome—appears at right as a succession of Cs, Ts, Gs, and As, representing the four nucleotides cytosine, thymine, guanine, and adenine.

recognition of the incalculable work that lies ahead.

Once the genome has been read, continues Watson, "we will be interpreting it a thousand years from now." And in the opinion of Francis Collins, who discovered the gene responsible for cystic fibrosis, the timetable for change will probably be slow enough that society will have time to work out solutions to genetics' knotty problems.

"It's conceivable," Collins says, speaking of genetic medicine in the early 1990s, "that we'll have effective treatments in the next ten to fifteen years, but I couldn't swear to it. We're just starting down this path, feeling our way in the dark. We have a small lantern in the form of a gene, but the lantern doesn't penetrate more than a couple of hundred feet. We don't know whether we're going to encounter chasms, rock walls, or mountain ranges along the way. We don't even know how long the path is."

CCCTGTGGAGCCACACCCTAGGGTTGGCCA
ATCTACTCCCAGGAGCAGGGAGGGCAGGAG
CCAGGGCTGGGCATAAAAGTCAGGGCAGAG
CCATCTATTGCTTACATTTGCTTCTGACAC
AACTGTGTTCACTAGCAACTCAAACAGACA
CCATGGTGCACCTGACTCCTGAGGAGAAGT
CTGCCGTTACTGCCCTGTGGGGCAAGGTGA
ACGTGGATGAAGTTGGTGGTGAGGCCCTGG
GCAGGTTGGTATCAAGGTTACAAGACAGGT
TTAAGGAGACCAATAGAAACTGGGCATGTG
GAGACAGAGAAGACTCTTGGGTTTCTGATA
GGCACTGACTCTCTCTGCCTATTGGTCTAT
TTTCCCACCCTTAGGCTGCTGGTGGTCTAC
CCTTGGACCCAGAGGTTCTTTGAGTCCTTT
GGGGATCTGTCCACTCCTGATGCTGTTATG
GGCAACCCTAAGGTGAAGGCTCATGGCAAG
AAAGTGCTCGGTGCCTTTAGTGATGGCCTG
GCTCACCTGGACAACCTCAAGGGCACCTTT
GCCACACTGAGTGAGCTGCACTGTGACAAG
CTGCACGTGGATCCTGAGAACTTCAGGGTG
AGTCTATGGGACCCTTGATGTTTTCTTTCC
CCTTCTTTTCTATGGTTAAGTTCATGTCAT
AGGAAGGGGAGAAGTAACAGGGTACAGTTT
AGAATGGGAAACAGACGAATGATTGCATCA
GTGTGGAAGTCTCAGGATCGTTTTAGTTTC
TTTTATTTGCTGTTCATAACAATTGTTTTC
TTTTGTTTAATTCTTGCTTTCTTTTTTTTT
CTTCTCCGCAATTTTTACTATTATACTTAA
TGCCTTAACATTGTGTATAACAAAAGGAAA
TATCTCTGAGATACATTAAGTAACTTAAAA
AAAAACTTTACACAGTCTGCCTAGTACATT
ACTATTTGGAATATATGTGTGCTTATTTGC
ATATTCATAATCTCCCTACTTTATTTTCTT
TTATTTTTAATTGATACATAATCATTATAC
ATATTTATGGGTTAAAGTGTAATGTTTTAA
TATGTGTACACATATTGACCAAATCAGGGT
AATTTTGCATTTGTAATTTTAAAAAATGCT
TTCTTCTTTTAATATACTTTTTTGTTTATC
TTATTTCTAATACTTTCCCTAATCTCTTTC
TTTCAGGGCAATAATGATACAATGTATCAT
GCCTCTTTGCACCATTCTAAAGAATAACAG
TGATAATTTCTGGGTTAAGGCAATAGCAAT
ATTTCTGCATATAAATATTTCTGCATATAA
ATTGTAACTGATGTAAGAGGTTTCATATTG
CTAATAGCAGCTACAATCCAGCTACCATTC
TGCTTTTATTTTATGGTTGGGATAAGGCTG
GATTATTCTGAGTCCAAGCTAGGCCCTTTT
GCTAATCATGTTCATACCTCTTATCTTCCT
CCCACAGCTCCTGGGCAACGTGCTGGTCTG
TGTGCTGGCCCATCACTTTGGCAAAGAATT
CACCCCACCAGTGCAGGCTGCCTATCAGAA
AGTGGTGGCTGGTGTGGCTAATGCCCTGGC
CCACAAGTATCACTAAGCTCGCTTTCTTGC
TGTCCAATTTCTATTAAAGGTTCCTTTGTT
CCCTAAGTCCAACTACTAAACTGGGGGATA
TTATGAAGGGCCTTGAGCATCTGGATTCTG
CCTAATAAAAAACATTTATTTTCATTGCAA
TGATGTATTTAAATTATTTCTGAATATTTT
ACTAAAAAGGGAATGTGGGAGGTCAGTGCA
TTTAAAACATAAAGAAATGATGAGCTGTTC
AAACCTTGGGAAAATACACTATATCTTAAA
CTCCATGAAAGAAGGTGAGGCTGCAACCAG
CTAATGCACATTGGCAACAGCCCCTGATGC
CTATGCCTTATTCATCCCTCAGAAAAGGAT
TCTTGTAGAGGCTTGATTTGCAGGTTAAAG
TTTTGCTATGCTGTATTTTACATTACTTAT
TGTTTTAGCTGTCCTCATGAATGTCTTTTC

TOOLS FOR MAPPING GENES

Scientists engaged in the international project to identify and locate the approximately 100,000 genes carried on human chromosomes employ a number of mapping techniques that help them circle ever closer to their quarry. A human gene may be hundreds of thousands of base pairs long, and the slightest deviation in sequence can give rise to any one of more than 3,000 genetic diseases. To complicate the picture further, only some 5 percent of the three billion or so base pairs in the human genome actually make up genes; the function of the rest is a mystery.

Finding a particular gene, then—and any tiny glitch in it—is a task akin to trying to locate a particular blond child on Earth from an observation post on the moon. As illustrated on the next several pages, geneticists must, in effect, determine the continent before they can zero in on the neighborhood. Genetic linkage maps, which use studies of inherited traits, can sometimes narrow the search to a general location on a given chromosome. So-called contig maps (from "contiguous") put large chunks of DNA into physical order on the chromosome. A third technique, called sequencing, gets down to determining the order of the bases, pair by pair.

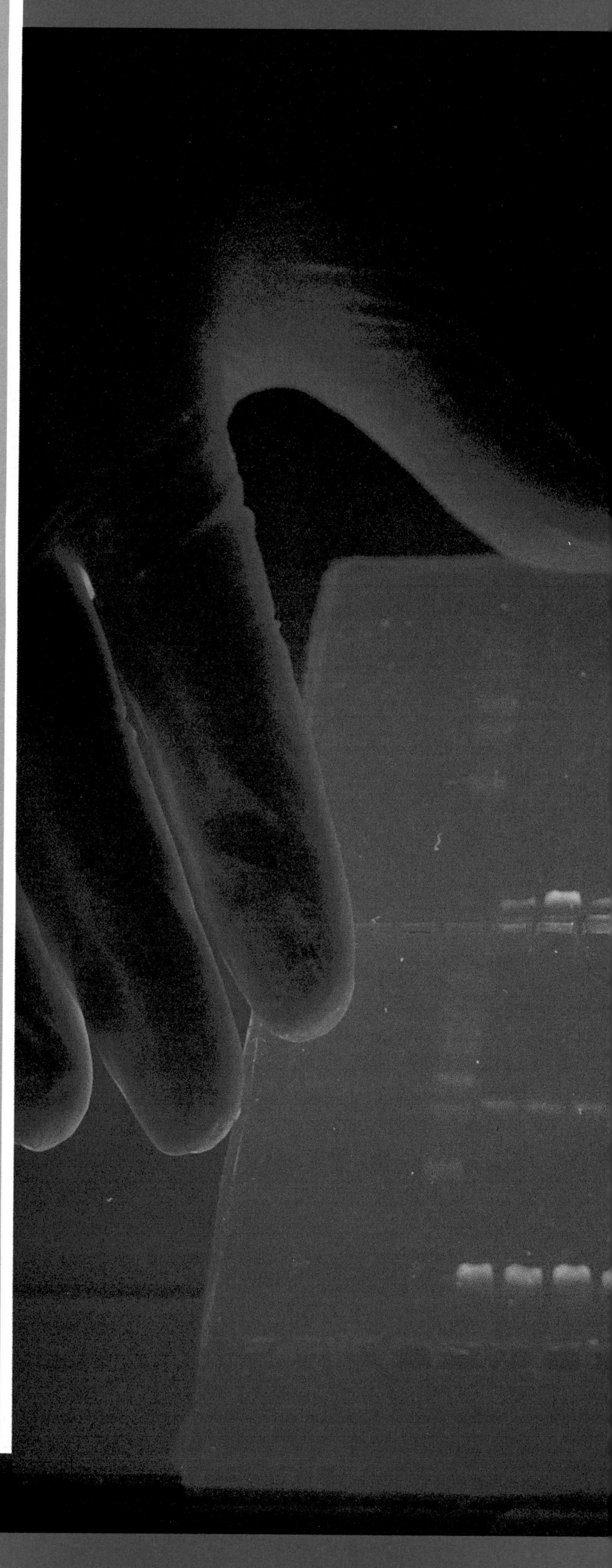

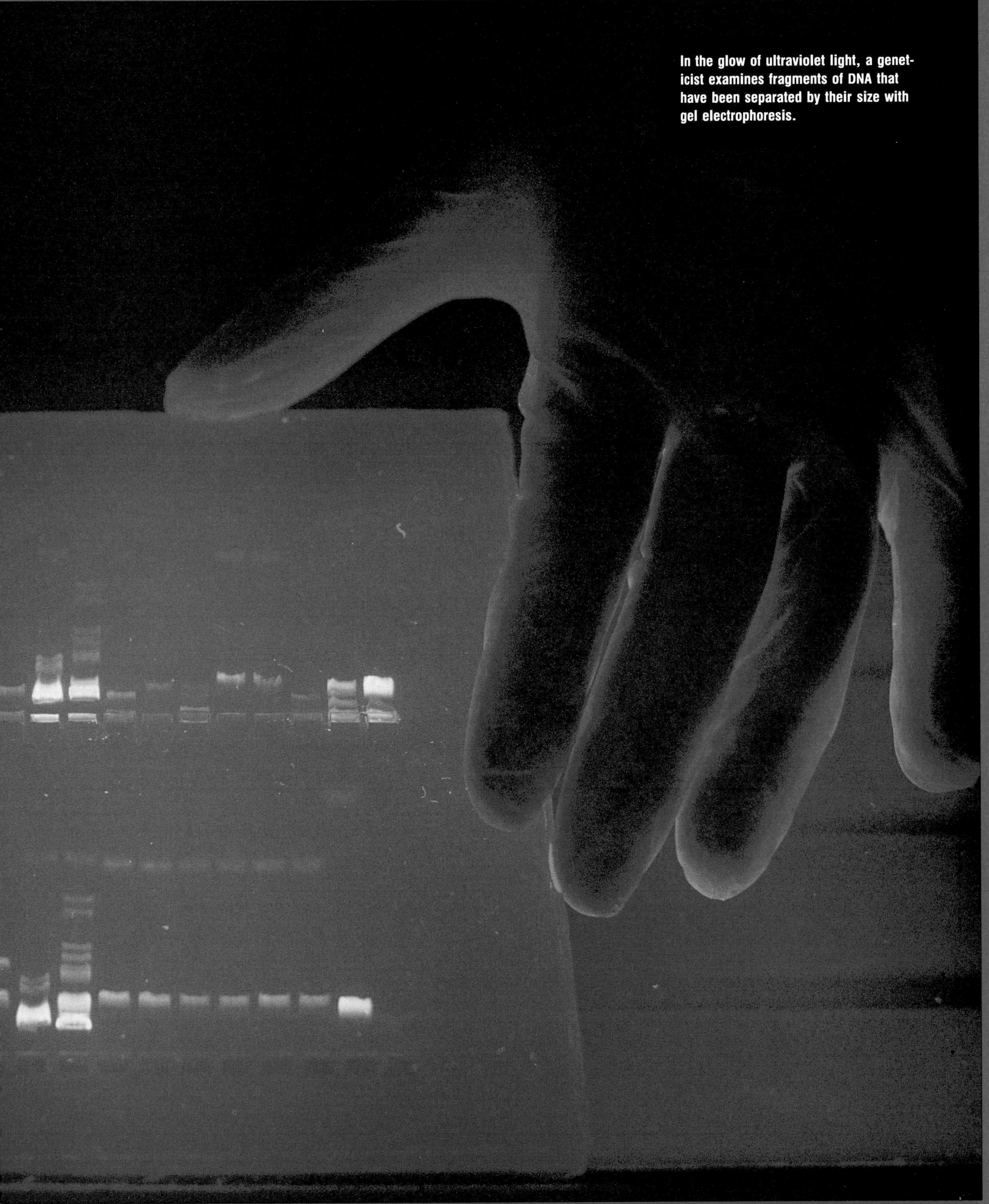

In the glow of ultraviolet light, a geneticist examines fragments of DNA that have been separated by their size with gel electrophoresis.

LINKING TRAITS TO CHROMOSOMES

Identifying genes on a chromosome by means of a linkage map is a matter of inference. Scientists have observed that certain DNA characteristics are almost always inherited with a particular disease or trait, implying that the genes involved are somehow linked. This linkage results from one aspect of meiosis (*pages* 23-25), in which members of a pair of chromosomes can cross over and exchange portions of DNA with each other. The nearer two segments of DNA are to each other, the more likely they will travel together during this exchange.

To that end, researchers have identified a number of DNA markers—attributes that vary slightly among individuals and tend to run in families. One type of marker, restriction fragment length polymorphisms, or RFLPs, can be identified using proteins called restriction enzymes, which cut a DNA strand at sites where the enzyme recognizes a sequence—CCTGAGG, for example; because the sequences can occur in different places, the process thus generates different-length DNA fragments. A sorting procedure (*pages* 77-79) leads to the formation of characteristic patterns, which show up in an image called an autoradiograph. By comparing the autoradiographs of healthy people with those of people who have an inherited disease, scientists can often identify a marker that occurs only in people with the disease. Since researchers use probes of known location to create autoradiographs, they can determine the general region on the chromosome where the disease-causing mutant gene must lie.

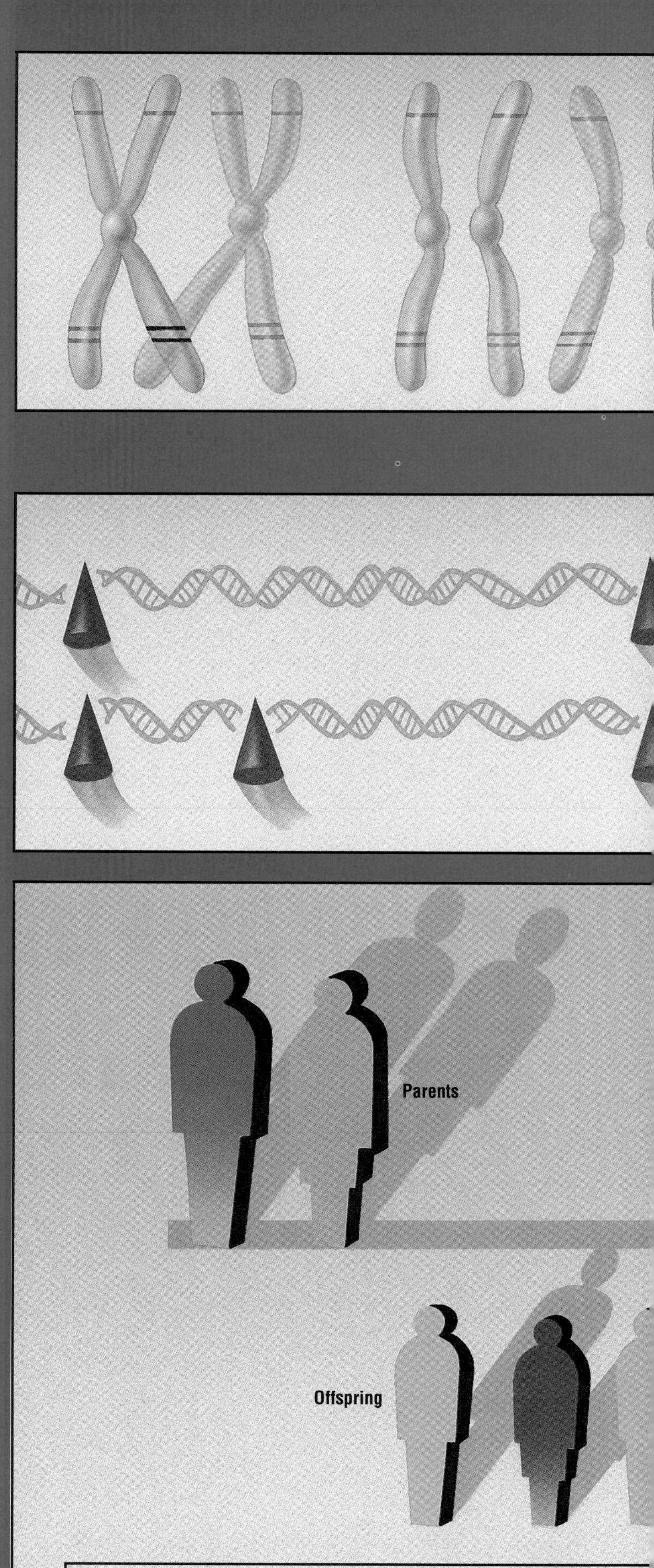

A MEASURE OF PROXIMITY. During meiosis, chromosomes that will ultimately become part of a new egg or sperm cell pair up and exchange portions of their DNA *(far left).* The three lines in each chromosome represent either a marker or a gene; the two that lie near one another are less likely to be separated during the crossing over, while the marker at the other end of the chromosome is very likely to travel independently of either of the other two. Linkage maps rely on such recombinations in order to determine links between mutant genes and DNA markers.

MARKING THE SPOT. To isolate a marker associated with a disease-causing gene, researchers use restriction enzymes *(red cones),* which cleave DNA at sites where the enzyme recognizes a specific sequence of nucleotides. In a strand of DNA from a healthy person *(top),* the enzymes identify two sites. But a strand taken from an afflicted person *(bottom)* contains an additional cutting site. Because this new site is inherited with the disease, it is assumed to lie near the disease-causing gene on the chromosome.

ESTABLISHING A LINK. To determine if a given hereditary disease is associated with an identifiable marker, researchers take DNA samples from families afflicted by the disease. Each sample is cut with various restriction enzymes and an autoradiograph is produced. The autoradiograph below reveals that in this family, the DNA from the father, a daughter, and two sons, all of whom have the disease, contains both versions of a marker that has two alleles; disease-free family members have one.

A LINKAGE MAP. In this simplified linkage map for the X chromosome, arrows point to the regions where genes for various diseases are believed to lie. The X chromosome is currently the one that researchers know most about, because X-linked inheritance is easily studied in males; indeed, the locations of some of the diseases noted here have been identified. For other chromosomes, however, a linkage map can give only the general location and the relative order of traits along the chromosome; maps detailing actual physical distances require techniques such as contig maps *(pages 92-93)* and sequencing *(pages 94-95).*

ARRIVING AT A MORE PRECISE ORDER

Linkage maps using RFLPs and other markers can bring scientists only to within roughly five million base pairs of a gene. One type of physical map, called a contig map, increases the level of detail down to about 500,000 base pairs by, in effect, taking chromosomes apart and then putting them back together.

The pieces of the genetic jigsaw puzzle are created by exposing many copies of each chromosome to a number of different restriction enzymes. Each of these assorted fragments of human DNA is then attached to the essential parts of a yeast chromosome for purposes of mass production of the human DNA. The resulting yeast artificial chromosomes, or YACs, are then cut by restriction enzymes to release the human DNA.

Because the process began with up to the equivalent of five genomes, and because each copy of each chromosome was cut in different places by different enzymes to produce DNA fragments of various lengths, the fragments of human DNA in YACs will tend to overlap. The YAC fragments, meanwhile, are screened to detect whether they carry special markers called sequence-tagged sites, or STSs. Through careful analysis, scientists can effectively reassemble the pieces of a chromosome in proper order (*far right*)—with, thanks to the STSs, many points along the chromosome tagged and cataloged.

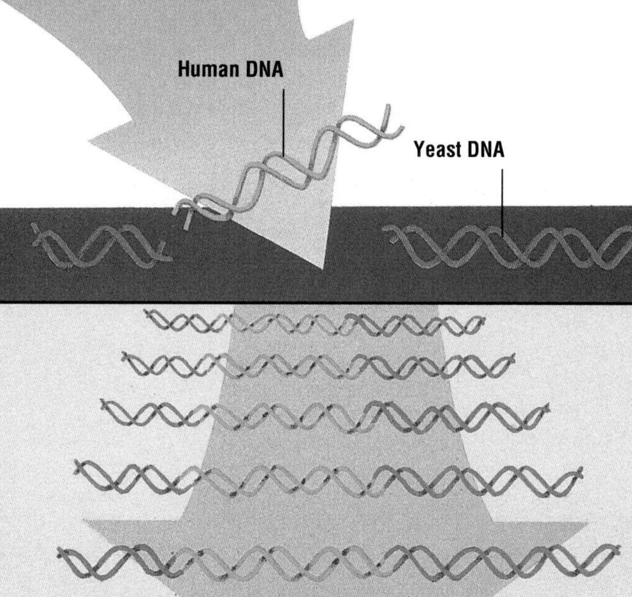

MAKING GENETIC COPIES. To make enough large pieces of human DNA for study, fragments resulting from the action of restriction enzymes are attached to a yeast chromosome. As yeast cells copy their own DNA *(orange)*, they also copy the human DNA *(blue)*, producing a so-called yeast artificial chromosome, or YAC. YACs are screened to identify the markers they contain.

TESTING FOR MARKERS. Step 1. To determine which markers lie on a given YAC, the yeast DNA is divided among several containers and mixed with a solution of all four nucleotides, a growth-catalyzing enzyme, and specific primers—short DNA segments with a sequence identical to the first few bases of a particular STS marker or its complement. (Here, the markers are represented by five symbols.) The solution is then heated, causing the DNA to unravel into its two complementary strands.

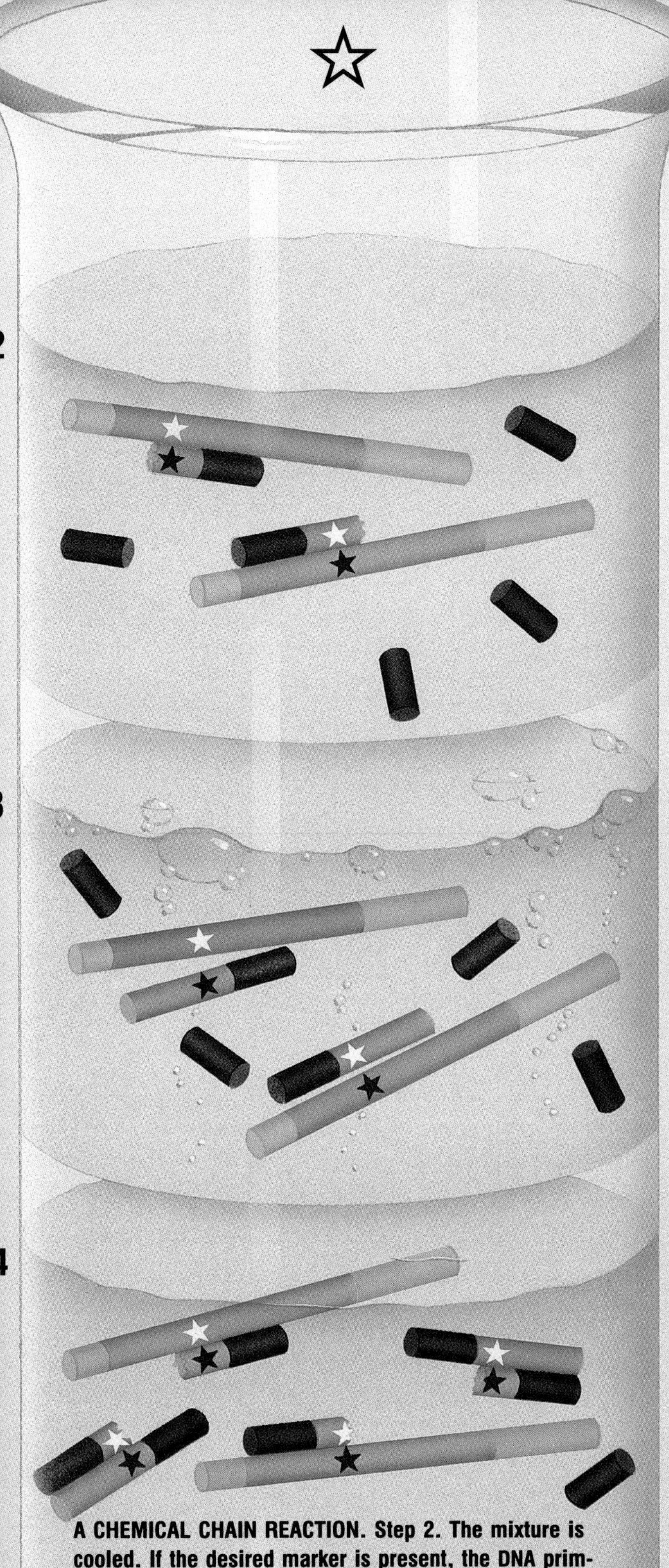

A CHEMICAL CHAIN REACTION. Step 2. The mixture is cooled. If the desired marker is present, the DNA primers will attach, each to its complementary strand, and start formation of a new strand that includes the marker (denoted, in this case, by black stars and white stars). Step 3. The mixture is reheated, splitting the original marker DNA segments and their newly formed complementary strands. Step 4. Cooling allows both new and old strands to serve as templates in yet another round of synthesis. With repeated heating and cooling cycles, the marker DNA fragment can be copied exponentially.

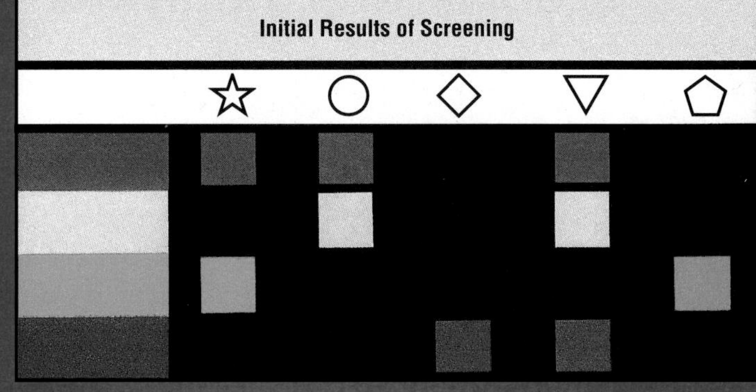

At the end of the replication process illustrated at left, researchers can tell which fragments of DNA contain which markers. In this example, the strand of DNA was chopped into four fragments—represented by four colors—and the fragments were screened for five markers. Scientists know that any fragments sharing one or more markers must overlap, but until they do further analysis, they still do not know the relative order of these fragments in the chromosome itself.

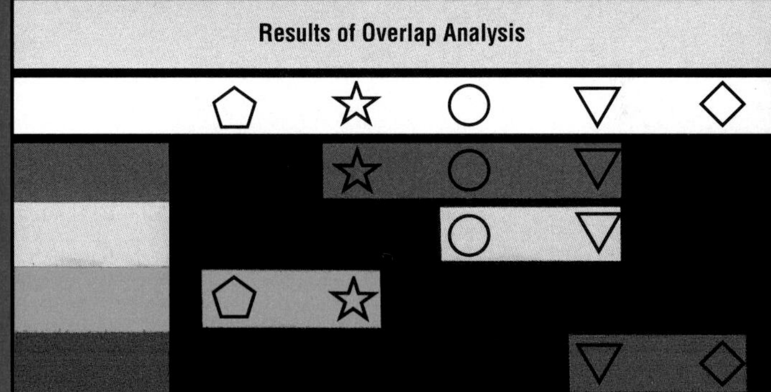

Assisted by computers, researchers begin to put the markers in order. In effect, the analysis compares fragments against one another. The red fragment, for example, contains the diamond and triangle, while the yellow fragment holds the triangle and the circle. The triangle must be an overlap point for the two fragments, with the diamond and the circle on either side. Then, since the blue fragment contains the star in addition to the circle and triangle, the relative order of those three markers becomes apparent—leaving the pentagon and the diamond at opposite ends.

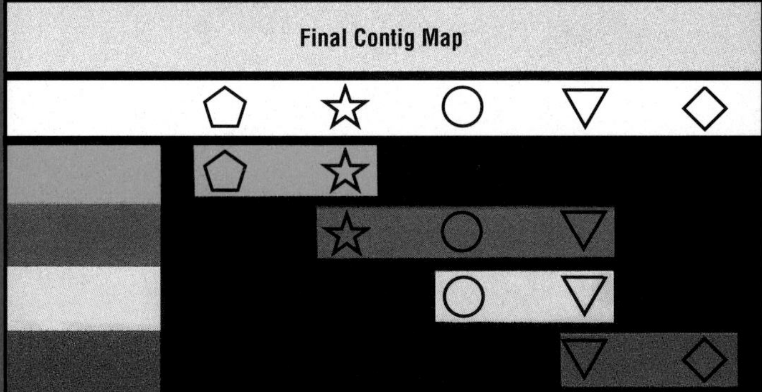

With the marker sequence in place, the proper fragment sequence (green, blue, yellow, red) is revealed. Moreover, knowing the approximate length of each fragment gives researchers an idea of how far apart the markers are. For example, if the green fragment is 250,000 base pairs long, the pentagon and star markers on that fragment can be no more than that many bases apart—much finer resolution than is possible with a linkage map.

HIGH-RESOLUTION GENE MAPPING

To map the human genome in its entirety requires finding the specific order of all the bases on each chromosome—although scientists are concentrating first on determining sequences of the small percentage of base pairs that make up genes. Limited sequencing has been done all along, of course, to create markers such as sequence-tagged sites, but to sequence the entire genome has only become really feasible with the advent of sophisticated computers.

The process is similar to making a contig map, but instead of starting a chain reaction on both DNA strands, scientists generate complementary copies of very small fragments—perhaps 400 bases long—of one strand. Moreover, during each cycle of replication, formation of the complementary fragment is randomly halted by the attachment of a special base, or terminator. This produces strands of varying lengths, each ending in an identifiable base. When the strands are ordered by length, the sequence of the terminators reveals the sequence of its complement (*far right*).

If all scientists have to go on is a rough idea of a gene's location (from linkage maps, say), they zero in on it by first constructing contig maps and then looking for special sequences known to signal where genes begin or end. As in all forms of gene mapping, of course, researchers must contend with large segments of noncoding DNA. And in any event, locations identified by one method must be confirmed by other methods before researchers can go forward with gene therapy efforts (*pages* 96-105).

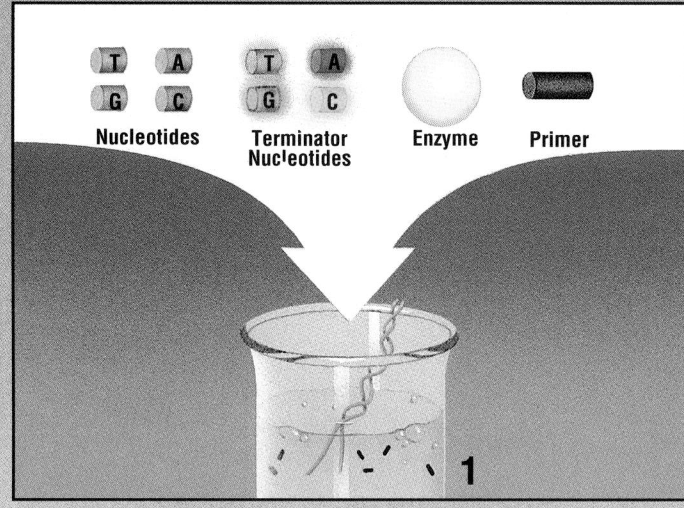

PREPARING THE DNA. Step 1. A fragment of DNA is mixed with growth enzymes; a supply of all four nucleotides for building more DNA; a primer that corresponds to the first few base pairs of a specific segment of the DNA fragment; and a supply of terminators—chemically altered versions of the four nucleotides that, once attached to a DNA strand, will stop further growth. Each terminator is also tagged with a colored dye, one for each of the four nucleotides, that fluoresces when exposed to laser light. The solution is then heated to split the two DNA strands; only one strand serves as a template for the copying.

A CYCLICAL REACTION. Step 2. As the solution cools, the primer binds to its complementary DNA fragment. Step 3. The enzyme adds bases from the solution to the primer-generated strand until a terminator base—randomly drawn from the bases floating in the soup—attaches to the strand, halting further growth. Step 4. The solution is heated again, causing the newly formed DNA to separate from its template. Step 5. The solution cools, and the formation of complementary DNA begins again. After repeated cycles of heating and cooling, the random attachment of the terminators will have created strands of every possible length.

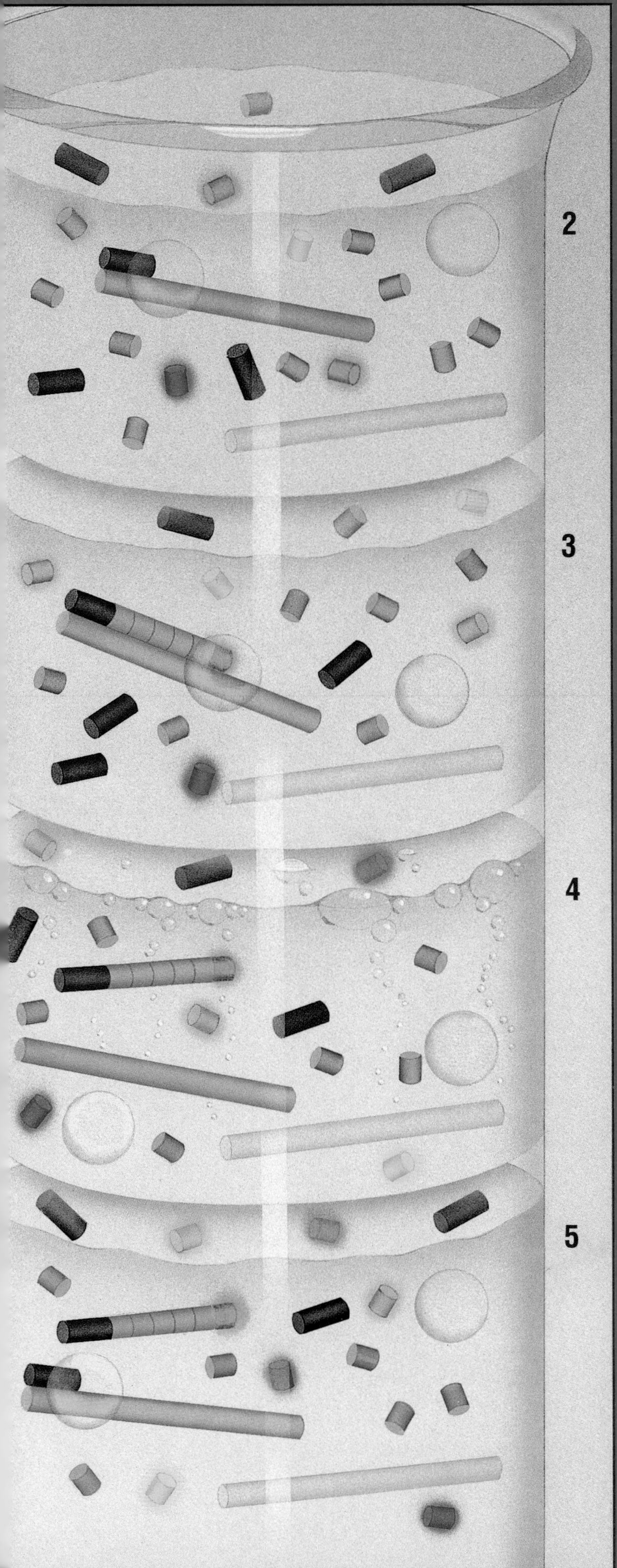

2

3

4

5

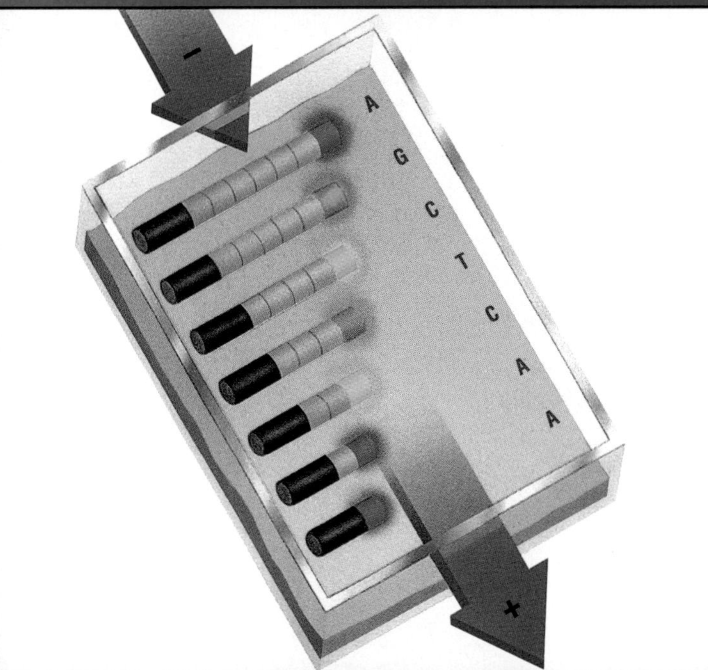

READING THE SEQUENCE. The newly formed complementary fragments of DNA are subjected to gel electrophoresis *(page 79)*, which sorts the strands by size, the smaller pieces traveling faster toward the positively charged end of the gel plate. Ranging in length from one base beyond the primer to the 400-or-so base length of the entire original fragment, each piece ends with a specific terminator base. In this example, the original fragment consisted of the primer and seven bases, resulting in seven different lengths. Exposed to a laser, each fragment's terminator fluoresces in its characteristic color, identifying the sequence of terminators, in this case the sequence AACTCGA.

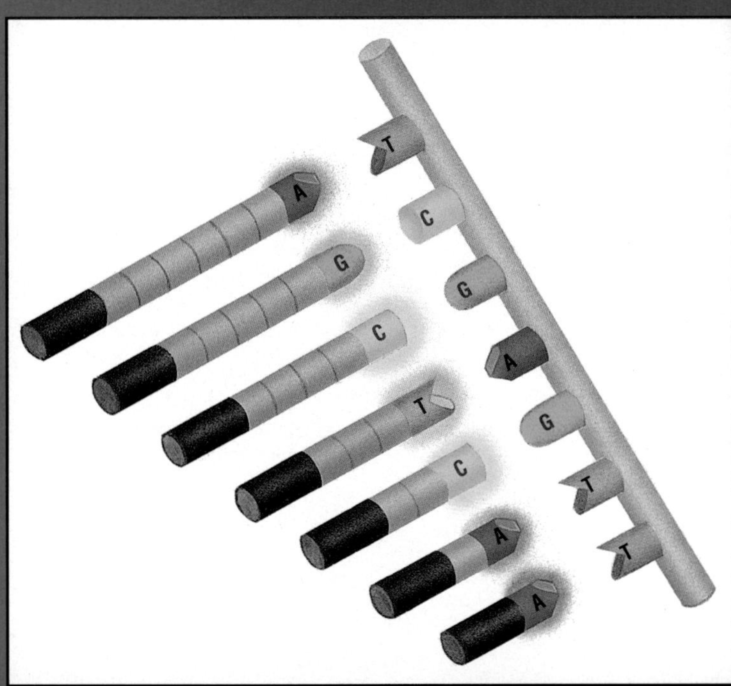

THE FINAL RESULT. Because the sequence of terminators is complementary to the sequence of bases on the original fragment of DNA, the original sequence is revealed: from AACTCGA comes TTGAGCT.

DESIGNER GENES FOR BETTER HEALTH

Genetic researchers have long sought practical applications for their growing understanding of the building blocks of life. In laboratories such as the one shown here, they are putting their knowledge to work to tackle a variety of diseases, employing several different strategies that hinge on carefully designed alterations to genetic coding. The first targets have been disorders caused by a single defective gene, but gene therapy—widely acclaimed as one of the most advanced weapons in modern medicine's arsenal—may eventually help treat conditions whose origins are less narrowly defined, such as heart disease or AIDS.

Gene therapy involves the insertion of new genetic material into cells in a patient's body, to correct or enhance a particular cellular function, to increase the vulnerability of diseased cells, or in some cases to block their operation altogether. Puzzling out the best approach to a specific disease is only part of the problem; researchers must also find safe, efficient vehicles—called vectors—to carry the new DNA to targeted cells.

Vectors already in use include chemical solutions and synthetic fat molecules, as well as specially modified viruses. Because of their natural ability to infect living cells (*pages* 70-73), viruses make excellent vectors—provided their own genetic instructions have first been adjusted to render them harmless.

Gene therapists must be alert to certain dangers in their work, such as the possibility that the integration of viral DNA into a cell's genome might accidentally activate so-called oncogenes—specialized genes that can trigger the uncontrolled cellular growth of cancer. But contrary to a widely held misconception, the standard forms of genetic manipulation present no danger at all for future generations, because they do not target the sex cells that pass along genetic traits. Indeed, gene therapy's main impact on the future may be to bring an end to the devastating toll of genetic diseases.

SUPPLYING A HEALTHY GENE

Because of a variety of key attributes, viruses with DNA rather than RNA at their core are sometimes preferred as carriers for therapeutic genes. For example, some of these viruses do not require that a cell be dividing in order to infect it—a feature that enables the viruses to home in on specific classes of cells within the body, such as those of the muscles, liver, and lungs. Furthermore, the genome of many DNA viruses does not actually integrate with chromosomes in target cells, but exists independently within the nucleus. The cellular genetic material thus remains intact, and there is less

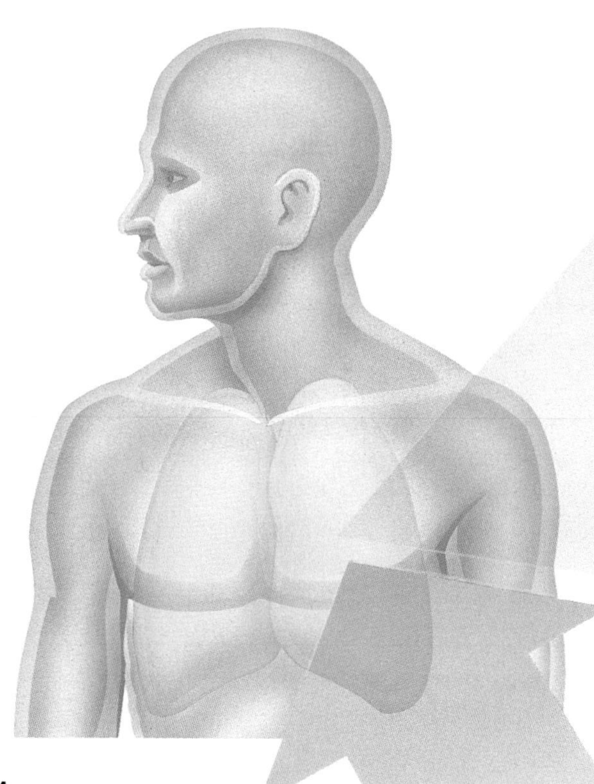

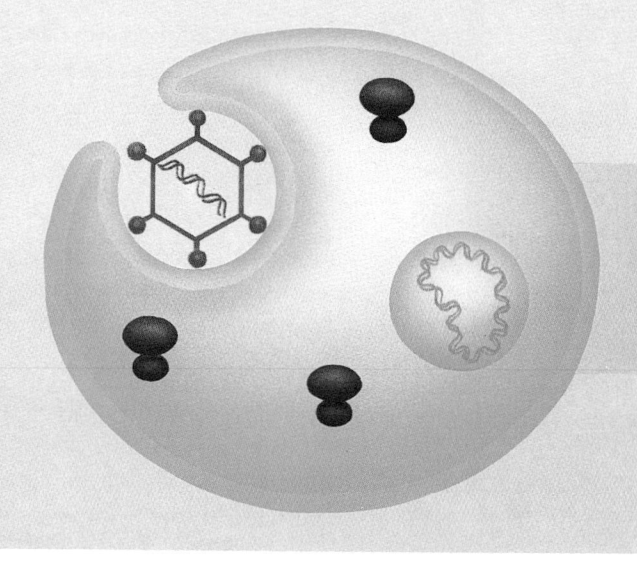

2 After the virus attaches to a lung cell's receptors, the cell membrane begins to envelop it, forming a vesicle, or small cavity, in the cytoplasm inside the cell, which also contains ribosomes *(purple)* and the nucleus with its own DNA *(blue)*.

1 Modified adenoviruses *(red hexagons, below right)* carry a gene *(red double helix)* for the protein missing in cystic fibrosis victims. Fibers projecting from the viruses match receptors on the surface of cells in the patient's lungs, allowing the viruses to attach there. The altered viruses are delivered to the lungs through a flexible tube inserted through the mouth or nose.

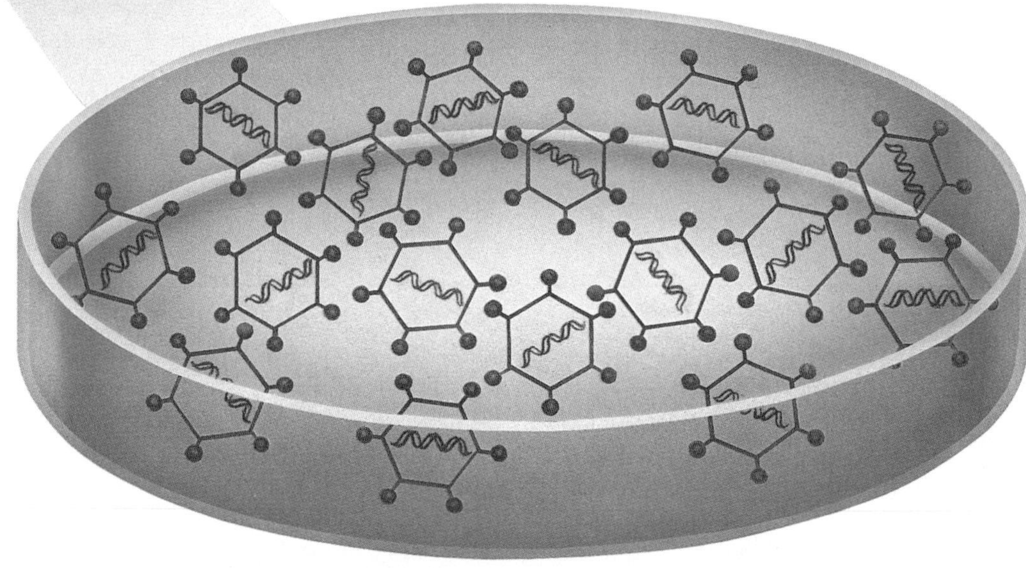

chance of the viral DNA activating a cancer-causing oncogene.

Among the most promising DNA vectors is one of the viruses that cause the common cold—the adenovirus, shown on these pages in its role in treating cystic fibrosis (CF). Chosen because of its affinity for epithelial cells that line the airways of the lungs, the adenovirus is modified with a specialized protein-synthesis gene. This protein, CFTR, absent in the lung cells of patients with CF, helps prevent the damaging buildup of mucus in the lungs that characterizes the disease.

Although offering hope to CF patients, adenovirus therapy does not amount to a permanent cure. The viruses have been altered so they cannot reproduce, and the cells they affect eventually grow old and die. To keep the treatment going, new batches of the therapeutic adenovirus must be administered every few months.

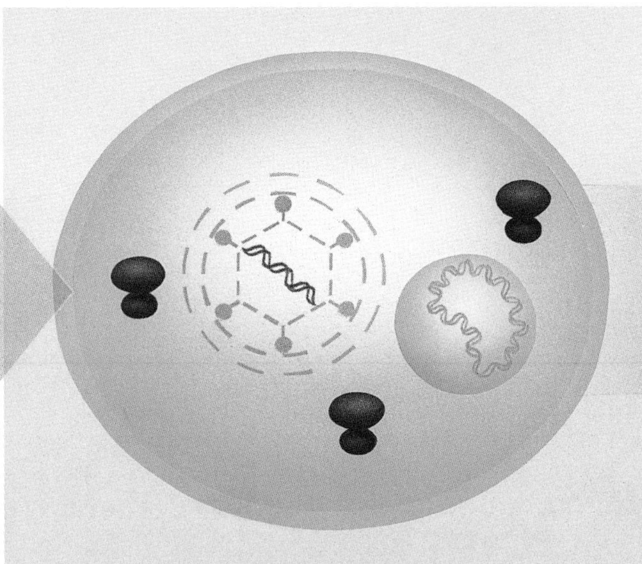

3 The vesicle has detached from the cell membrane and headed toward the nucleus. Cellular enzymes begin to break down the vesicle and outer coat of the virus, and the therapeutic viral DNA will soon be free to escape into the cytoplasm and then enter the nucleus.

4 In the nucleus, the viral DNA *(red)* remains distinct from the cellular DNA *(blue)*, directing all cellular function toward the production of therapeutic proteins. Transcription of the viral DNA yields single strands of messenger RNA *(red)*, which migrate to the cytoplasm.

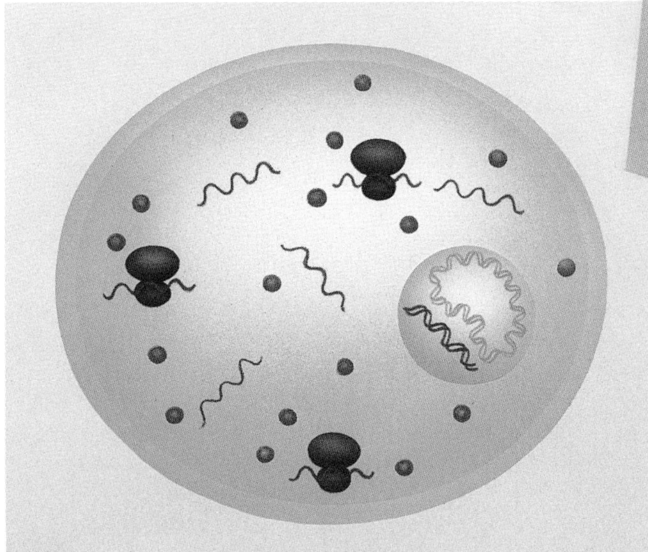

5 The cell's ribosomes translate the messenger RNA to synthesize the missing protein *(red spheres)*. All lung cells that receive the new therapeutic gene will continue to produce the protein until they die off in a few months.

TAGGING CELLS FOR ATTACK

While some viral vectors carry genes that help correct malfunctioning cells, others bear a more deadly genetic cargo, designed to eliminate defective cells altogether. In the therapeutic approach shown here, a modified retrovirus—that is, a virus containing RNA rather than DNA—is used to infect the rapidly dividing cells of an inoperable brain tumor. The retrovirus inserts a gene that codes for an enzyme associated with the herpesvirus; the presence of this enzyme makes the tumor cells susceptible to an antiviral drug called ganciclovir.

The retrovirus with this so-called

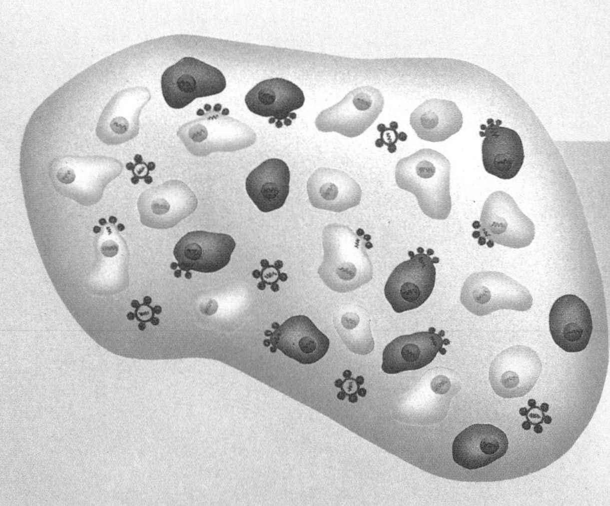

2 Inside the tumor, retroviruses *(spiked red circles)* bud from the mouse cells and begin attaching to tumor cells *(gray)*, bypassing the healthy brain cells *(blue)*.

1 Mouse cells *(below, right)* infected with a genetically altered retrovirus become factories for reproducing the virus. Budding retroviruses, each carrying the suicide gene, are preparing to detach from the cell membrane and infect other cells. The virus-infected mouse cells are injected into the patient's brain tumor *(gray mass, above)*.

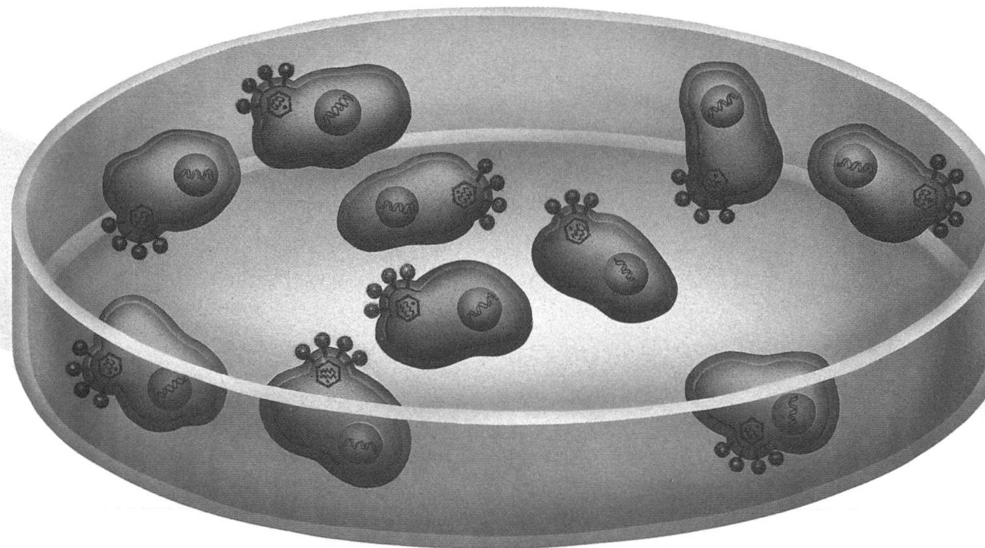

suicide gene cannot reproduce on its own, so it is packaged in specially engineered "helper cells" taken from the skin of mice. The genome of these helper cells provides genetic information that enables the cells to produce copies of the retrovirus (*pages* 72-73), which then can insert the therapeutic gene into multiple target cells.

Treatment begins with the injection of the mouse cells directly into the tumor. Because the retroviruses can combine their genetic material only with that of dividing cells, they have no effect on healthy brain tissue. In the proliferating tumor cells, however, they immediately go to work producing the herpes gene. The helper cells, meanwhile, go on generating retroviruses until the body rejects them, anywhere from two days to two weeks later. Any helpers that remain are eliminated along with the tumor cells when the ganciclovir is administered.

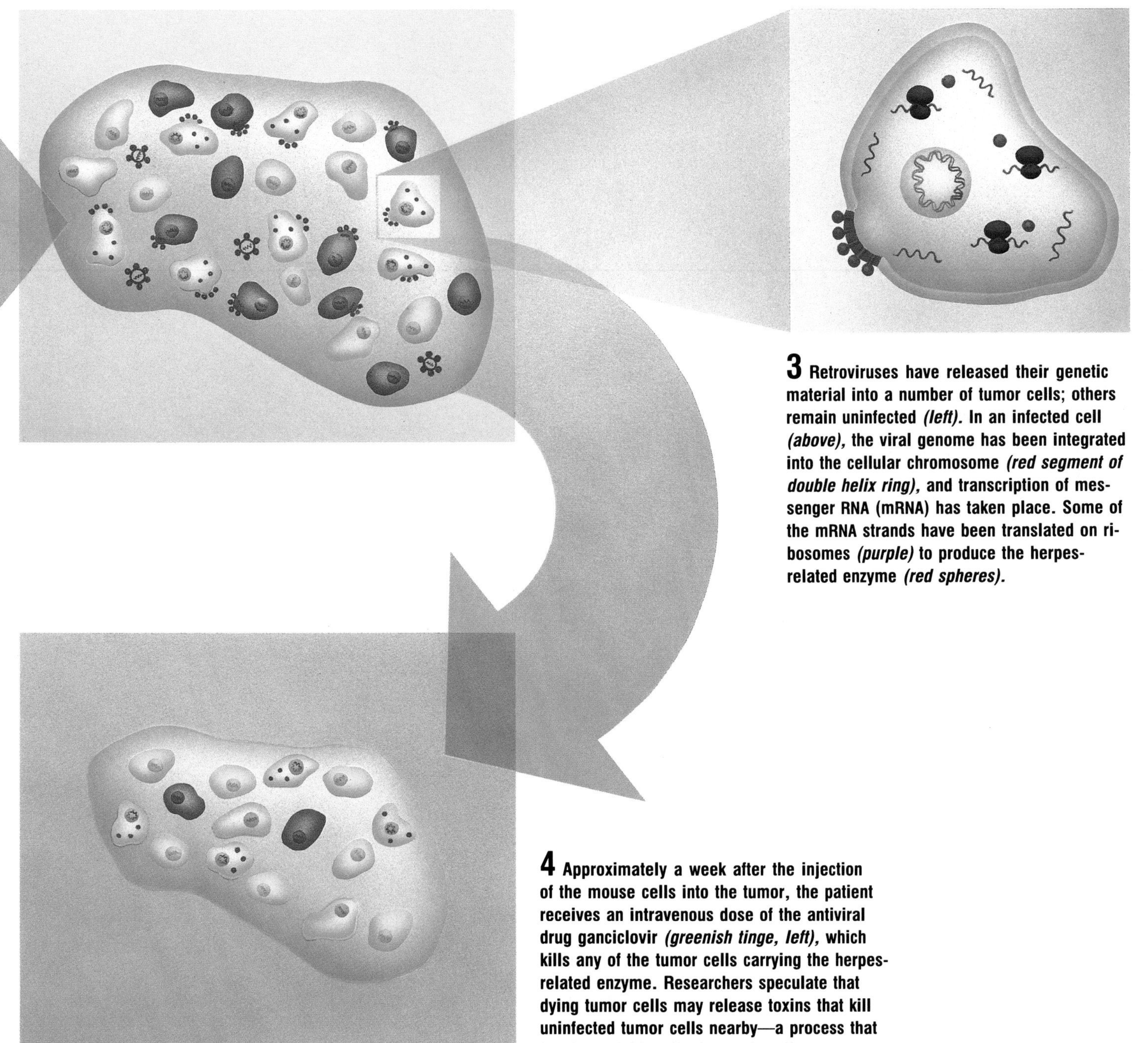

3 Retroviruses have released their genetic material into a number of tumor cells; others remain uninfected *(left).* In an infected cell *(above),* the viral genome has been integrated into the cellular chromosome *(red segment of double helix ring),* and transcription of messenger RNA (mRNA) has taken place. Some of the mRNA strands have been translated on ribosomes *(purple)* to produce the herpes-related enzyme *(red spheres).*

4 Approximately a week after the injection of the mouse cells into the tumor, the patient receives an intravenous dose of the antiviral drug ganciclovir *(greenish tinge, left),* which kills any of the tumor cells carrying the herpes-related enzyme. Researchers speculate that dying tumor cells may release toxins that kill uninfected tumor cells nearby—a process that has been dubbed the bystander effect.

A GENETIC CALL TO ARMS

In some experimental therapies, genetic material is used almost like a drug, being transplanted directly into the target tissue without the use of a viral vector. The technique illustrated below aims at malignant melanoma, a deadly form of skin cancer, using specially designed genes injected straight into the tumor. There they cause the production of a protein that activates the immune system against the cancerous cells.

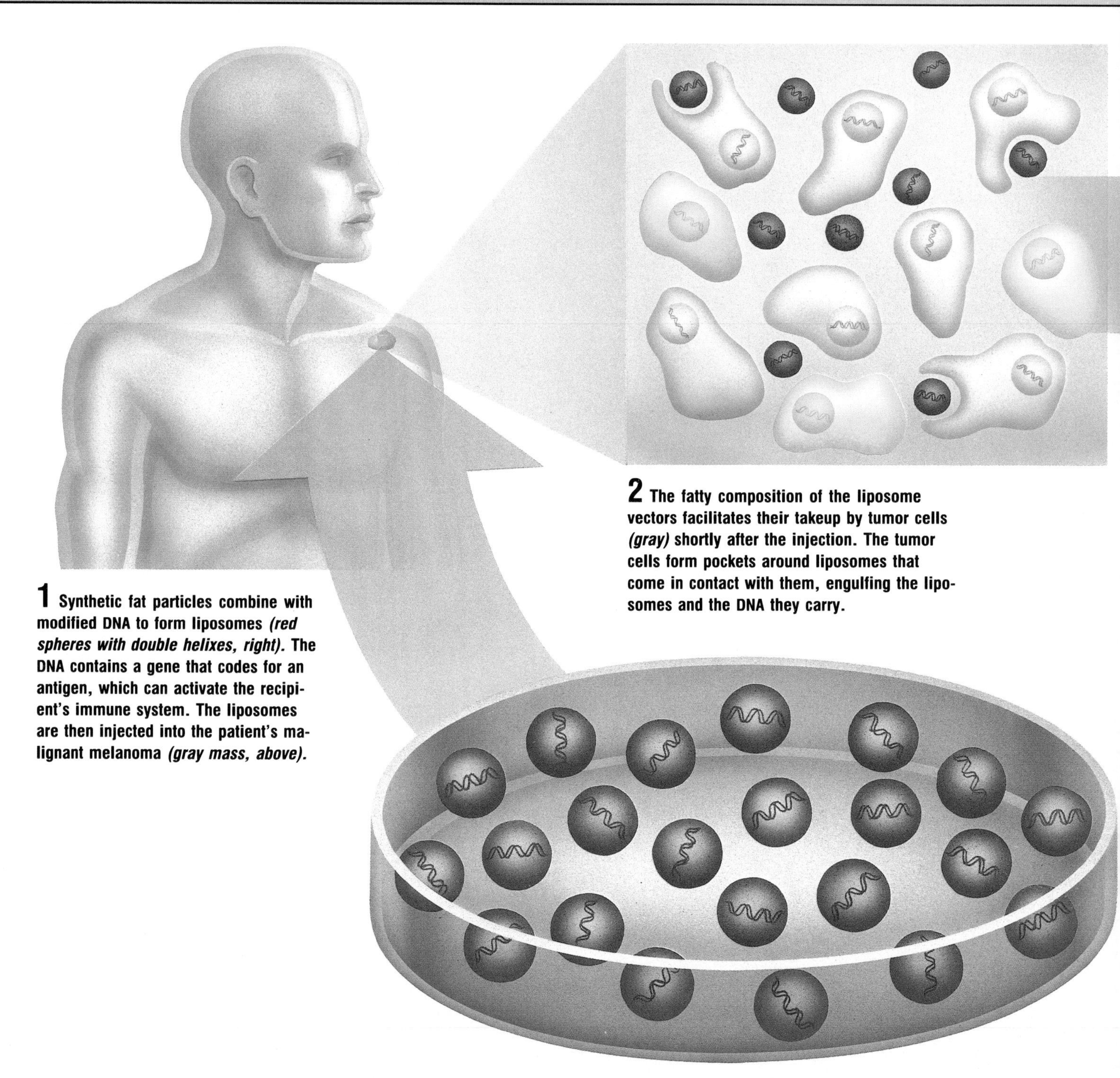

2 The fatty composition of the liposome vectors facilitates their takeup by tumor cells *(gray)* shortly after the injection. The tumor cells form pockets around liposomes that come in contact with them, engulfing the liposomes and the DNA they carry.

1 Synthetic fat particles combine with modified DNA to form liposomes *(red spheres with double helixes, right).* The DNA contains a gene that codes for an antigen, which can activate the recipient's immune system. The liposomes are then injected into the patient's malignant melanoma *(gray mass, above).*

Millions of copies of the gene are encapsulated in synthetic fat particles called liposomes, which protect the DNA and allow it to be ingested by tumor cells. Once inside, the genes cause the cells to generate an antigenic protein that is recognized by the body as a foreign substance. The alerted immune system promptly attacks all cells carrying the antigen and exterminates the tumor cells.

The liposome delivery system, although still experimental, could simplify DNA therapy and enable doctors to apply it to a broad range of diseases. The technique may also help make genetic approaches cheaper and more widely available.

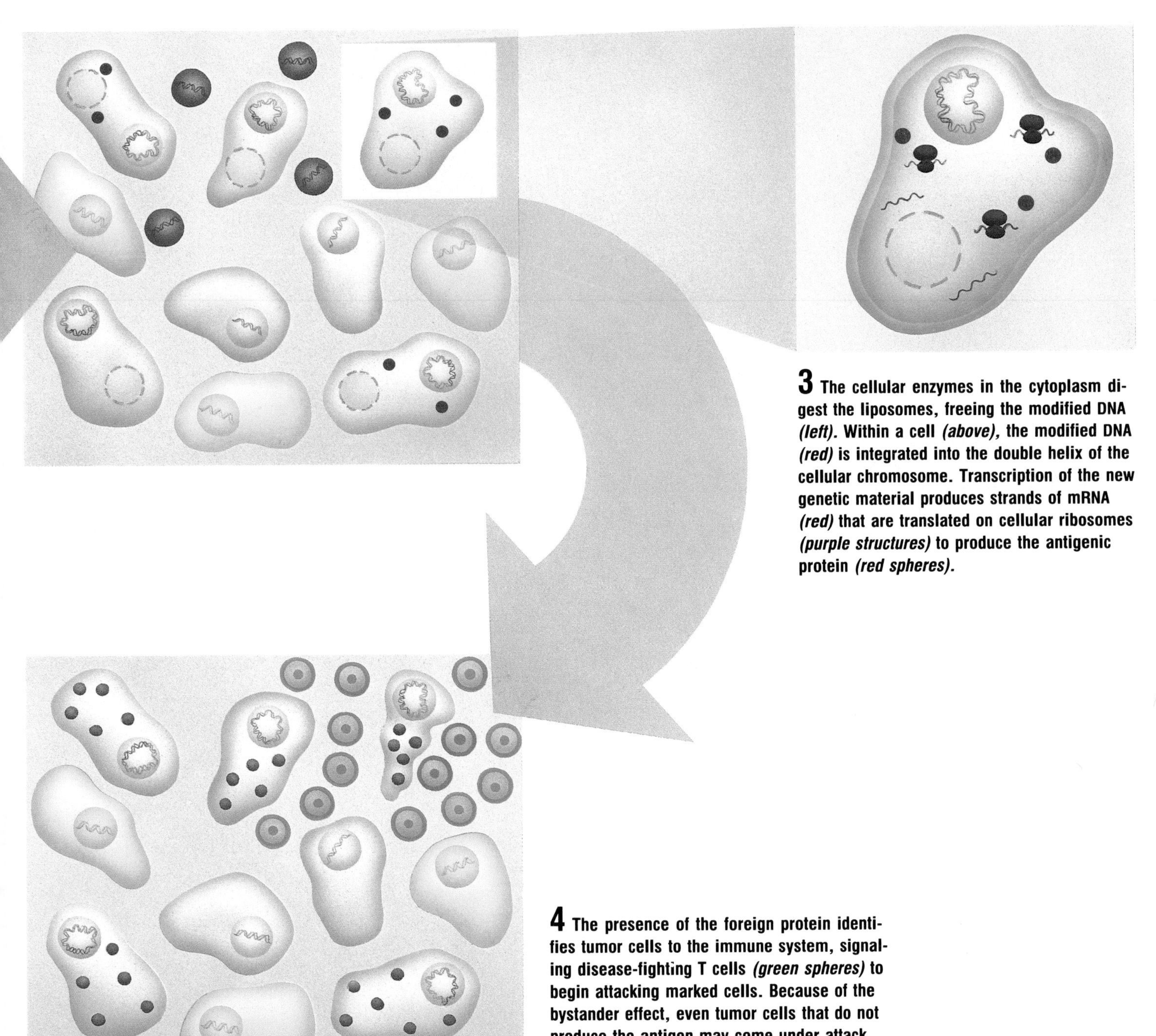

3 The cellular enzymes in the cytoplasm digest the liposomes, freeing the modified DNA *(left)*. Within a cell *(above)*, the modified DNA *(red)* is integrated into the double helix of the cellular chromosome. Transcription of the new genetic material produces strands of mRNA *(red)* that are translated on cellular ribosomes *(purple structures)* to produce the antigenic protein *(red spheres)*.

4 The presence of the foreign protein identifies tumor cells to the immune system, signaling disease-fighting T cells *(green spheres)* to begin attacking marked cells. Because of the bystander effect, even tumor cells that do not produce the antigen may come under attack.

BLOCKING A DEFECTIVE GENE

DNA therapy does not always involve inducing a cell to synthesize a therapeutic protein. With leukemia, for example, the method shown below works instead by blocking a defective gene before its instructions for protein synthesis can be carried out.

A bad gene, like every genetic sequence of nucleotides, has a specific message, or "sense," spelled out by its chemical subunits. For every such sequence there is also a complementary series, whose nucleotides spell out precisely the opposite message—the "antisense." This antisense strand is the only one that can bind firmly with the original genetic material.

Scientists can take advantage of this trait, designing DNA strands with

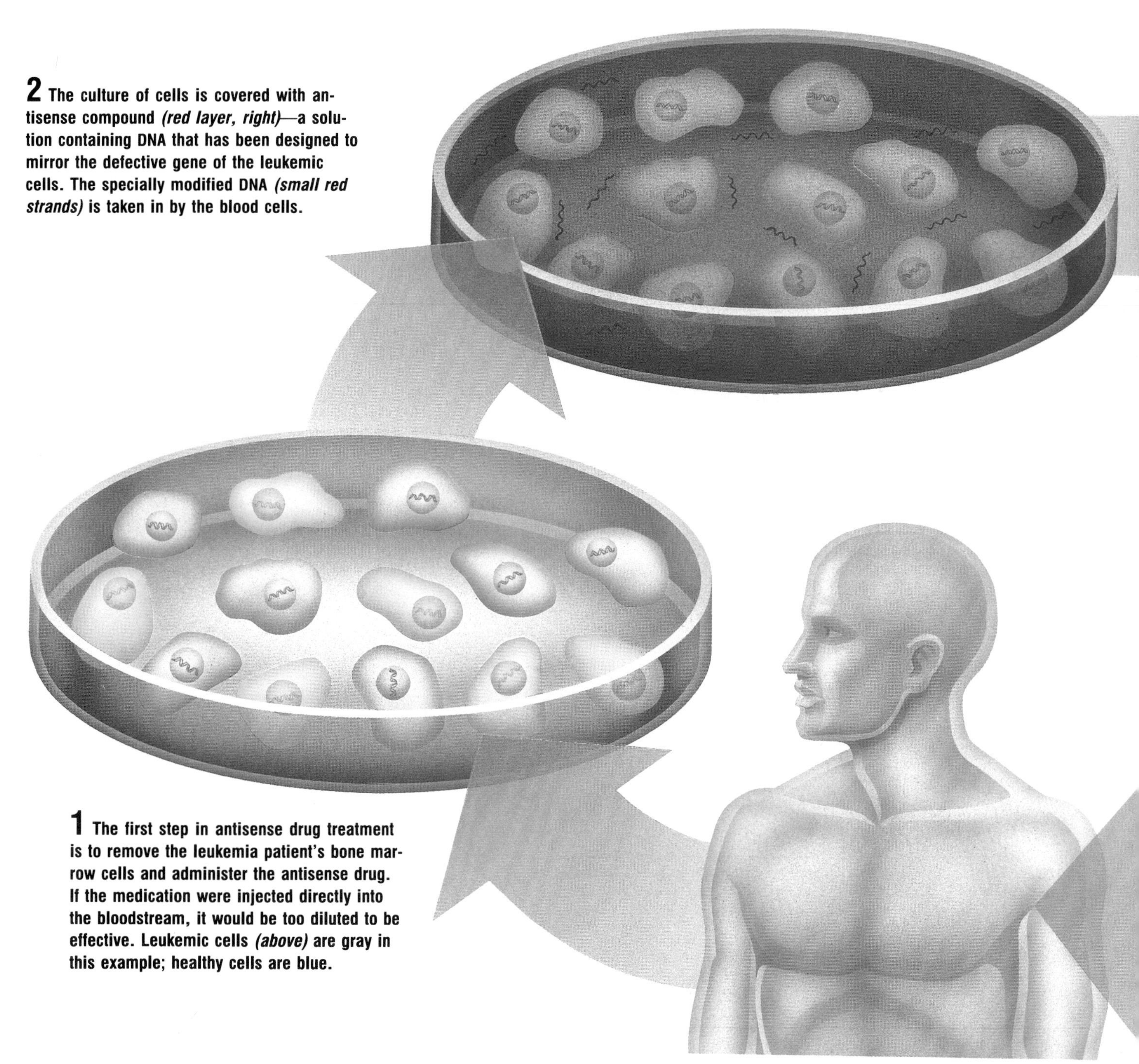

2 The culture of cells is covered with antisense compound *(red layer, right)*—a solution containing DNA that has been designed to mirror the defective gene of the leukemic cells. The specially modified DNA *(small red strands)* is taken in by the blood cells.

1 The first step in antisense drug treatment is to remove the leukemia patient's bone marrow cells and administer the antisense drug. If the medication were injected directly into the bloodstream, it would be too diluted to be effective. Leukemic cells *(above)* are gray in this example; healthy cells are blue.

the antisense of defective messenger RNA. When the antisense DNA binds with its target, the mRNA can no longer be translated for protein synthesis. The technique can destroy diseased cells by stopping the production of proteins essential to the cells' survival. Alternatively, the method can allow healthy genes to switch on by blocking expression of genes whose control of the cell function has a detrimental effect.

After first determining the sequence of the target gene, scientists synthesize a complementary antisense molecule, chemically modifying the sugar-phosphate backbone of the antisense sequence of DNA to enhance ingestion by the cell. The alteration also increases the resistance of the antisense strand to the destructive effects of cellular enzymes, thereby permitting the modified DNA to reach the cell nucleus.

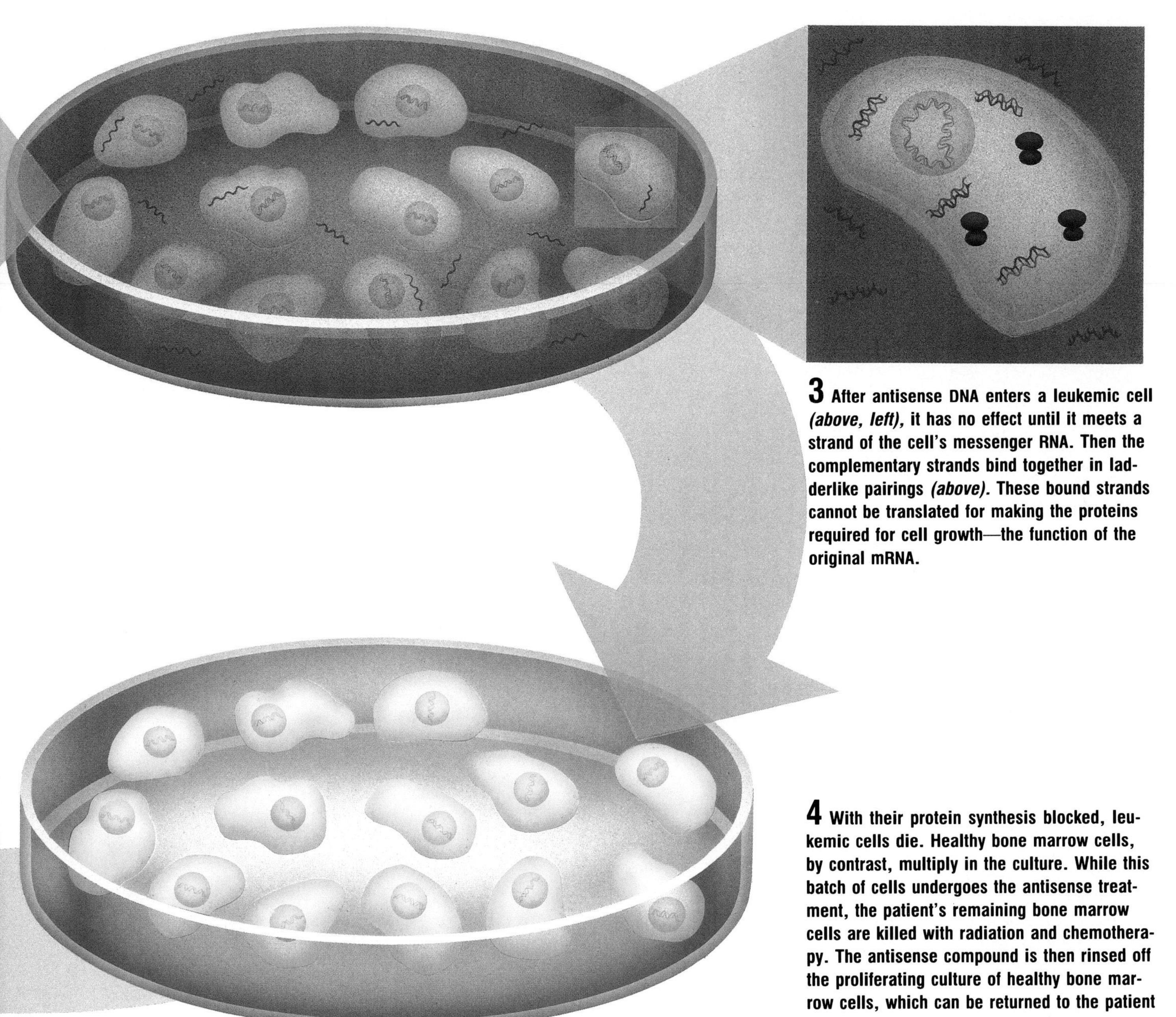

3 After antisense DNA enters a leukemic cell *(above, left),* it has no effect until it meets a strand of the cell's messenger RNA. Then the complementary strands bind together in ladderlike pairings *(above).* These bound strands cannot be translated for making the proteins required for cell growth—the function of the original mRNA.

4 With their protein synthesis blocked, leukemic cells die. Healthy bone marrow cells, by contrast, multiply in the culture. While this batch of cells undergoes the antisense treatment, the patient's remaining bone marrow cells are killed with radiation and chemotherapy. The antisense compound is then rinsed off the proliferating culture of healthy bone marrow cells, which can be returned to the patient to make blood that is free of leukemia.

3

The Question of Nature vs. Nurture

In the year 1827, a boy by the name of Francis Galton sat in his comfortable home near Birmingham, England, and penned a letter to his beloved older sister: "My dear Adele, I am four years old, and I can read any English book. I can say all the Latin substantives, adjectives and active verbs, besides fifty-two lines of Latin poetry. I can cast up any sum in addition and multiply by 2, 3, 4, 5, 6, 7, 8, 9, 10, 11. I read French a little and I know the clock."

Little Francis obviously had good reason to feel proud. Learning was almost like breathing to him, and it would remain so throughout his life. Over the next eight decades, he would make contributions in fields ranging from meteorology to physiology, earning an illustrious scientific reputation and innumerable honors, including a knighthood. Had he attempted any final summation of accomplishments like the list he had drawn up for Adele at age four, it would have been a lengthy list indeed.

One question held particular fascination for him: Do the characteristics and talents of people arise mainly from physical inheritance, or are they more a consequence of upbringing and other environmental factors? (He liked to frame the issue in the simpler words of

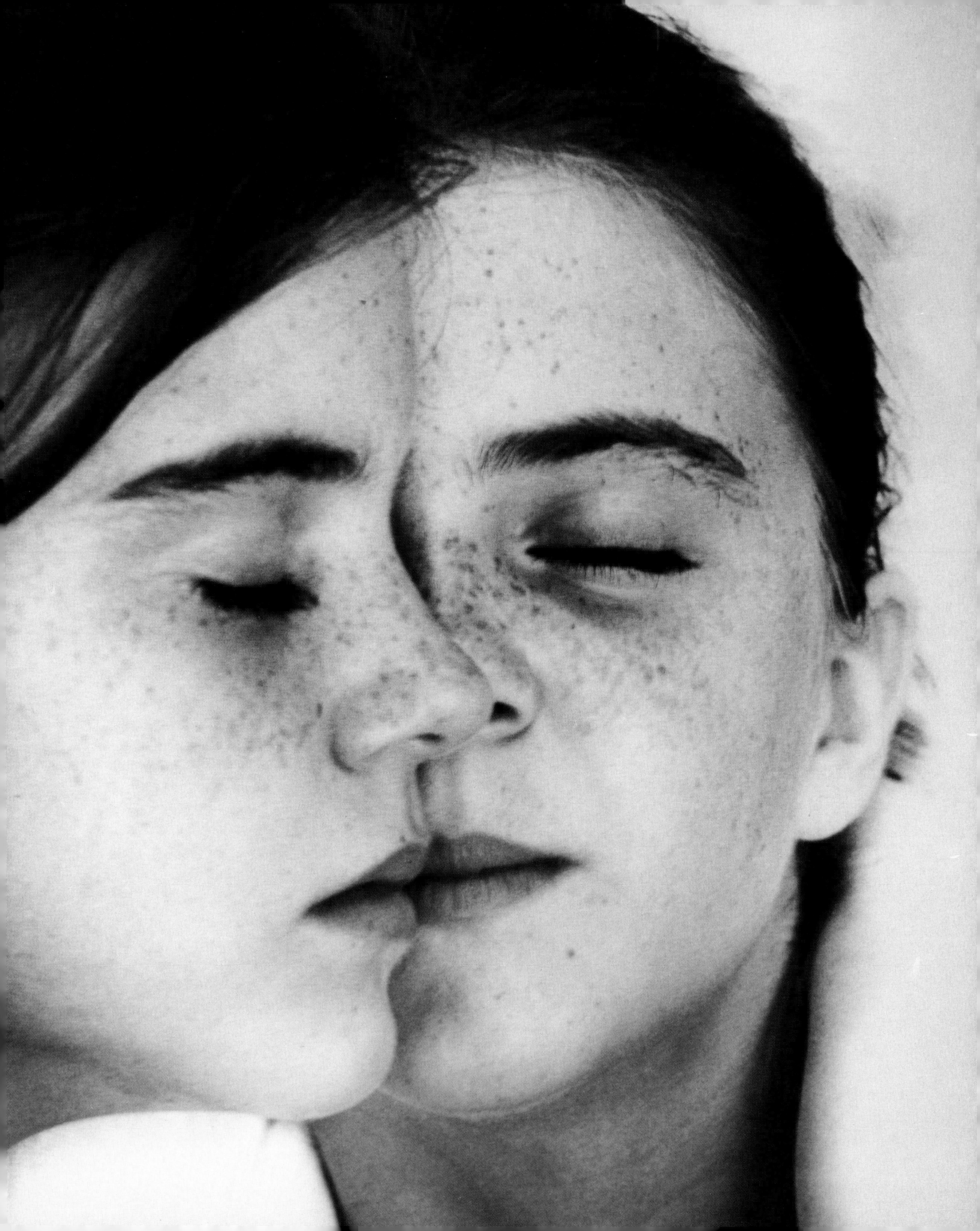

PRECEDING PAGE: So much alike that they could almost be reflections of one another, these twin sisters are perfect genetic duplicates, the result of a fertilized egg splitting after conception. Cases of identical twins raised apart provide a powerful research tool, clarifying the relative contributions of genes and the environment to human variability.

nature and *nurture*, terms that have been popularly associated with the debate ever since.) Although by no means the first person to ponder the matter, he was the first to pursue it scientifically, establishing a field of inquiry that would ultimately address many features of human behavior—among them cognition, language, personality, and susceptibility to such ills as alcoholism and schizophrenia. Some investigators have even looked for a genetic component underlying the fine details of everyday life, including likes and dislikes in attire, food, and so on.

Galton's principal interest was in discovering the basis of intelligence, and he brought an understandable bias to the problem: His own family—securely positioned in Britain's economic and social elite—seemed to offer proof that genius and bloodlines were linked. Among his many gifted relatives were his grandfather Erasmus Darwin, an eminent physician, philosopher, poet, and naturalist, and his cousin Charles Darwin, who attained scientific glory when he published his theory of evolution in 1859.

Such gifted clans were not uncommon. Galton was acquainted with a number of families that had made a disproportionate contribution to Britain's intellectual life, and he knew of comparable cases in other cultures and time periods—the painters Hans Holbein the Elder and Hans Holbein the Younger in 15th- and 16th-century Germany, for instance, or the great 18th-century German composer Johann Sebastian Bach, among whose numerous offspring were his three famous composer sons, Johann Christian, Wilhelm Friedmann, and Carl Philip Emanuel. (Musicality among the Bachs, in fact, extended well beyond those two acclaimed generations: Taking into account various forebears and descendants, the Bach line of professional musicians stretched across a chronological span of almost 200 years, from the 17th well into the 19th century.)

With a passion for quantification that characterized all of his work, Francis Galton undertook studies of marriages, social rank, and demographics, searching for statistical patterns that might bear on the question of inherited talent. His investigations convinced him that intellect, cultural achievement, and economic standing were indeed concentrated in certain families and that they had a hereditary component. Within a given social stratum, he said, "There is no escape from the conclusion that nature prevails enormously over nurture." But he was not satisfied with mere description of what he regarded as the facts. Sir Francis envisioned a branch of science that might someday allow humans to control the course of their own evolution. His name for this futuristic enterprise was *eugenics*, from the Greek for "well born."

At the turn of the century, Galton's ideas were taken up by a variety of thinkers who hewed to the optimistic philosophy of progressivism, believing that the march of human history rose ever upward toward the goal of perfection. The notion that the human line could be improved through selective breeding, like a line of cattle or dogs, dovetailed with prevailing attitudes about the innate superiority of certain races and classes. According to the logic of the eugenicists, if the most educated, talented, and wealthy people intermarried and bred exclusively with others of their kind, the offspring of their unions would be indubitably above the common run.

Ultimately, the goal of improving the human gene pool led to its most horrific manifestation in the Nazi gas chambers, the product of Adolf Hitler's dream of a superrace of pure Aryans. In the United States, eugenicists condemned such monstrous schemes, yet they advocated the equally pernicious view that, over time, reproduction among the poor and among people of color would, along with racial mixing, result in "inferior" individuals and trigger a decline in society's overall intellectual vigor, a downward trend that would eventually tail off into imbecilism.

Because of its profoundly negative associations, eugenics became widely discredited after the Second World War, supplanted by the countervailing school of thought known as behaviorism, which had been founded a few decades earlier by an American psychologist named John B. Watson. Behaviorism at its purest held that the effects of the environment—a child's upbringing and tutelage, social milieu, and degree of exposure to books, music, museums, and other cultural refinements—so far outweighed those of genes as to render the influence of heredity negligible. Watson maintained that it was within the power of educators to mold individuals at will, regardless of their parentage. The only differences between people in one profession and another were circumstantial, he believed, and social gaps could be closed by proper conditioning. "Give me a dozen healthy infants, well formed, and my own special world to bring them up in," Watson wrote in a ringing declaration of his faith in human malleability, "and I'll guarantee to take any one at random and train him to become any type of specialist I might select—doctor, lawyer, artist, merchant-chief, and yes, even beggar and thief, regardless of his talents, penchants, tendencies, abilities, vocations, and race of his ancestors."

Behaviorism remained ascendant through the 1950s and 1960s, largely on the strength of experiments that demonstrated a capacity for learning in such laboratory animals as white rats and pigeons. But animal experiments undermined some of the stronger behaviorist claims in the end, when it became clear that much animal learning was constrained and shaped by a particular species' genetic endowment. The pendulum thus began to swing back toward the position held by Sir Francis Galton—although researchers now knew vastly more about the workings of heredity than he did and had little interest in his notions of human improvement.

As investigators were increasingly realizing, just getting a handle on the basic facts of nature and nurture was challenge enough: Quite clearly, no simple recipe could account for human behavior. During the past four decades, in fact, the burgeoning field of behavioral genetics has relied heavily on statistical analysis to explore the relative—and complex—

EFFECTS OF THE ENVIRONMENT
MANKOFF
IDENTICAL TWINS SEPARATED AT BIRTH.

influences of genes and the environment. In part, the research has proceeded through selective breeding experiments with animals, whose genotypes can be readily manipulated in much the same manner as those of Mendel's pea plants. Such an approach is impossible and unethical with human beings, of course, and scientists have had to work backward from cognitive or behavioral attributes, as revealed by batteries of tests, to presumed genotypic and environmental influences. Perhaps the most tantalizing clues have come from research projects focusing on identical twins—subjects whose 100-percent-shared genetic inheritance presents a kind of uniform background against which environmental influences on behavior stand out with special clarity.

Not surprisingly, the twins approach was pioneered by Francis Galton. The number of his subjects was small, but he felt that the results were conclusive, demonstrating a very strong link between the physical inheritance that individuals bring into the world and the kind of people they become. Indeed, Galton considered the results of his studies of identical twins so unequivocal that he worried about how the public might respond. When he published the findings in a paper entitled "The History of Twins, as a Criterion of the Relative Powers of Nature and Nurture," he confessed, "My only fear is that my evidence seems to prove too much and may be discredited on that account, as it seems contrary to all experience that nurture should go for so little."

Modern twin studies, although far more carefully constructed and executed than anything Galton attempted, sometimes appear to run the same danger. Certain studies, for example, have suggested that genes play a role in such behavioral characteristics as neatness, respect for authority, and vividness of imagination—personality traits that a behaviorist such as John Watson would have firmly ascribed to the environment. Other studies appear to show—contrary to expectation—that differences in familial environments do not tend to shape individual personalities in a significant or consistent way. Practitioners in the field are certainly no strangers to controversy. But even their toughest critics would concede that valuable insights have emerged from the statistical thickets of behavioral genetics. Although certainties are in short supply, human behavior is more and more perceived to have a genetic component. One researcher sums it up in a phrase whose simplicity Sir Francis Galton might have admired. Genes, he says, provide "a rough sketch of life." Environment, endlessly interacting with the genetic framework, does the rest.

One of the many miracles of the living world is the ability of nature to extract almost infinite diversity from a relatively small portion of life's total genetic instruction kit. Most genes dictate sameness rather than differences. Honed by eons of evolution, they constitute a blueprint for the essential molecular architecture that living creatures need—the biochemical basics that enable cells to function. Because genes devote most of their DNA to instructions for this fundamental architecture, the world's tens of millions of species, ranging in size from submicroscopic bacteria to the behemothian blue whale, are kin in more ways than not. The fruit fly *Drosophila melanogaster*, that favorite laboratory subject of geneticists, is a case in point. At the phenotypic level—that is, in terms of physical form—*Drosophila* and *Homo sapiens* appear to have in common virtually nothing. But at the genotypic level, at least in terms of the genes that have been examined so far, about 71 percent of *Drosophila*'s genes have human counterparts—that is, the two sets of genes have similar functions. Between humans and other mammals the gap

Despite the range of physical traits that can be seen among peoples around the world—illustrated here by differences in hair and height that delight a 10-year-old American traveler and a Kenyan Masai girl of about the same age—all human beings share at least 99.9 percent of their genes.

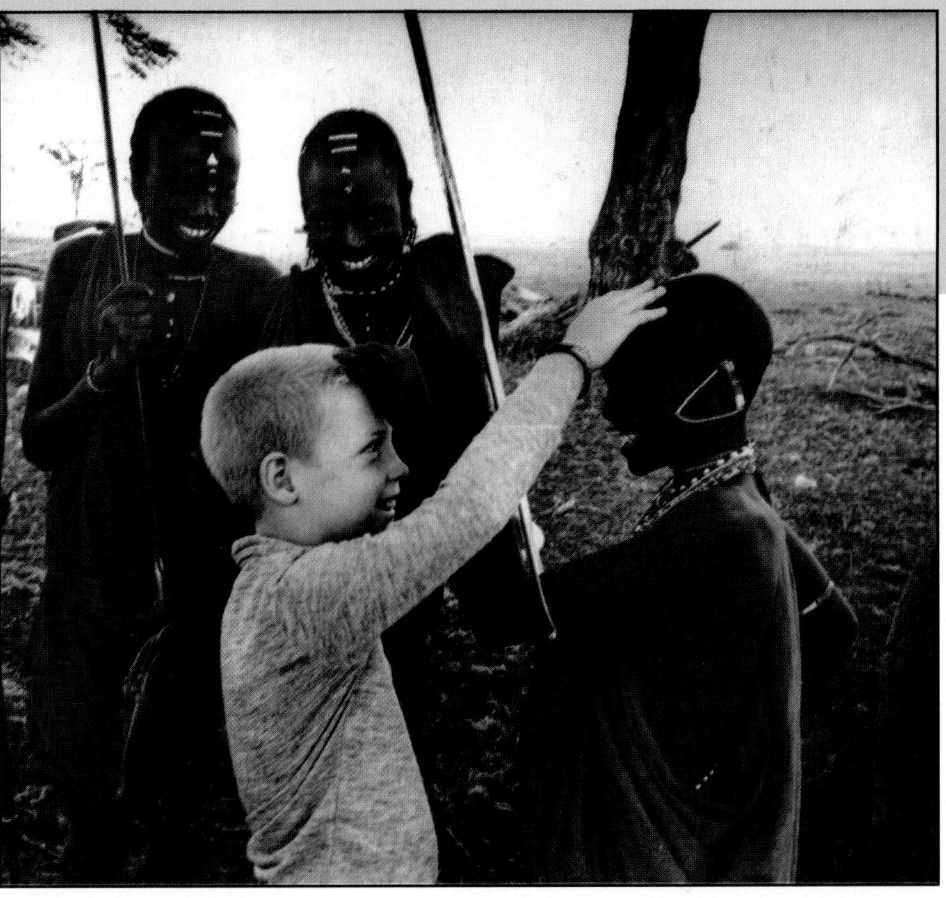

shrinks drastically: The genetic instructions that produce a human and those that produce a mouse differ by only 20 percent; a mere one percent separates human from chimpanzee.

For that matter, no representative of the human race boasts more than one-tenth of one percent variation in genetic material from his or her fellows—roughly three million base pairs of DNA out of a total of three billion. Yet from that tiny hereditary leeway have emerged individuals bearing a range of skin tones from ivory to ebony; eye colors from pink to charcoal; hair lanky to kinky. From that seemingly minimal allowance for innovation derive groups of people distinguished by their diminutive stature, like the Pygmy, or by their towering height, like the Watusi. The closer the relationship between individuals, the less the difference in base pairs in their DNA. Only about two million base pairs separate nonidentical siblings or parent and child. No genetic separation exists between identical twins, which account for three or four births out of every thousand, or between identical triplets or quadruplets, which are much rarer.

Yet regardless of what the genes prescribe, the environment will exert an influence on development, complicating the picture even for identical twins. The process begins early—almost instantly, in fact. As soon as a fertilized egg affixes to the uterine wall and begins growing, it becomes subject to external forces—what biologists call "developmental noise." The conditions within the womb, the diet of the mother, her level of stress, even the ambient sound level of her immediate surroundings can affect the progress of a fetus. If a pregnant woman smokes, drinks alcohol, falls prey to a virus such as German measles or the flu, or is exposed to toxic chemicals or excess radiation, the vulnerable fetus may suffer damage and the resulting infant display mental or physical aberrations.

Environmental effects multiply after birth. Compared with the outer world, the womb offers what are tantamount to controlled laboratory conditions. The emerging infant plunges into a chaos of light and sound, heat and motion. Every child experiences a distinct, individualized reality—a unique set of stimuli and social relations. Even children raised in the same household by the same caring parents will be exposed to different constraints—subtly different parental expectations, for example, or different degrees of parenting confidence and skill. Nurture, in the literal sense, can never be applied to any two individuals in exactly the same way.

From the moment of birth, that nurture is acting on a formidably complex organism. At one time, babies were regarded as a kind of living clay, lumpen masses born without vision or sensitivity, incapable of thought or emotions—in short, barely human. Such notions have been thoroughly demolished by psychologists; it is now known that infants possess remarkable skills very early on, including the ability to pay attention, to monitor sights, sounds, smells, heat and cold, even to tell human faces from other objects, and to recognize their mothers' voice. More important, the human genome conveys to a healthy child certain inalienable gifts, including spatial perception, a capacity for mimicry of expressions and gestures, the ability to remember, and certain fundaments of reasoning and logic. Some researchers believe the genes even provide the underlying structure of the human species' most distinctive attribute, language.

The word *infant* comes from the Latin *infans*, meaning "one who is unable to speak." In this label lies perhaps a partial justification for humankind's longstanding practice of underestimating babies: Being wordless, they are other than us, lacking that skill we count as quintessentially human. At the same time, babies soon display a powerful urge to talk—so clearly inborn that, in earlier eras, people sometimes wondered if actual words might not be inborn as well. For example, Akbar the Great, a 16th-century Mogul emperor of India, is supposed to have conducted an experiment in which he isolated a large group of infants in a house with silent nurses for a period of several years. The children grew up mute. In more recent instances, however, isolated children of normal mental capacity have learned to speak upon being given some training.

Clearly, the genotype does not encode specific words, and equally clearly, the ability to speak at all requires exposure to language relatively early on in life. Still, the genes may do more than simply produce an urge to speak; they may also dictate speech's very form.

The search for possible genetic underpinnings of language has been led by Noam Chomsky, a linguist at the Massachusetts Institute of Technology. Chomsky believes that all of the world's 6,000 different languages have a common structural core. Just as the human genome programs the embryo to grow human limbs and organs rather than those of a chicken or an alligator, so too it engineers a brain primed with what Chomsky calls a "language-forming" faculty. According to Chomsky, grammar, or the rules by which words are strung together in comprehensible arrangements, is not invented but instead arises from neurological circuitry shared across the human species. From this innate organizing system, which Chomsky calls a universal grammar, the mind is able to generate a seemingly infinite array of verbal statements that make sense. Children thus do not learn language from scratch but rather absorb vocabulary and plug the acquired words into an already existing mental framework that differentiates between nouns and verbs, objects and subjects, and automatically relates clauses one to another.

Some linguists have looked for clues to the hypothetical universal grammar in creole languages, which originate as makeshift tongues that enable people of markedly different cultures to communicate. One scholar who has investigated these collision-born languages is Derek Bickerton. While teaching at the University of Hawaii in the mid-1970s, he and a group of graduate students recorded hundreds of hours of local speech. Bickerton noted that Hawaiian "creole English," which arose in the late 19th century among immigrants from all over the world—including China,

WHERE TWINS DIFFER. Even when indistinguishable in all other physical regards, identical twins can often be told apart by their fingerprints, as illustrated by the boxed areas in the photograph below, showing different arrangements of ridges. One speculation is that environmental influences in the womb may irrevocably affect the development of these characteristic patterns.

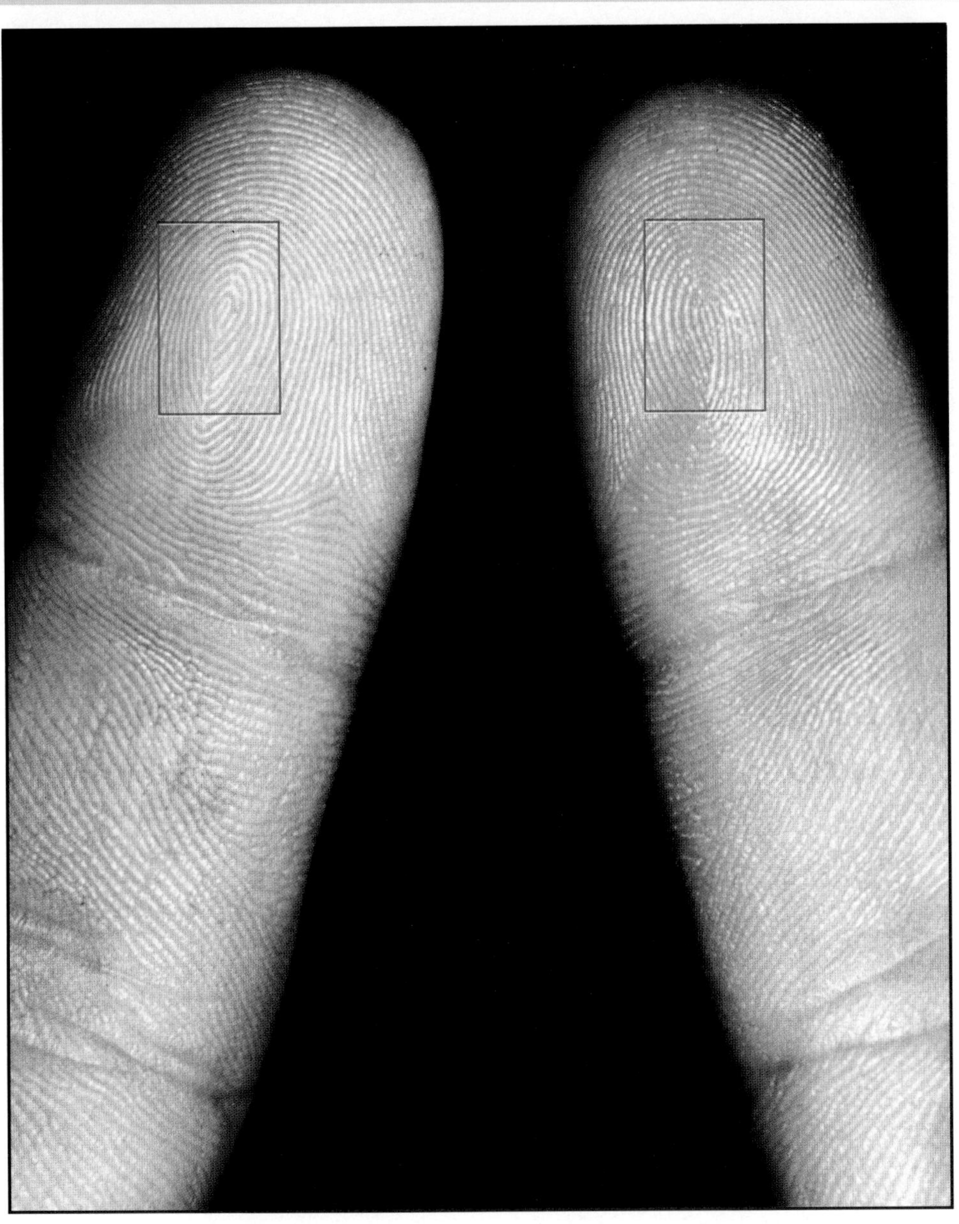

Japan, Portugal, Puerto Rico, the Philippines, and the United States—bore similarities to Caribbean creole languages that originated among Africans some two centuries earlier. His interest was piqued, and he went on to examine other creole languages around the world. The Hawaiian-Caribbean resemblances seemed to apply elsewhere.

Like the rudimentary language spoken by toddlers, creole tongues often appear to contravene the rules of accepted usage, yet, also like baby-speak, they have an internal consistency of their own, possibly because of some universal program writ in the genes—a "bioprogram," as Bickerton calls it. He points out, for example, that all creole languages make a clear distinction between actions that have been accomplished and those that are unaccomplished. A Jamaican indicating that a person had gone to wash would say either "Him gone for bathe" (that is, he went with the intention to bathe or wash) or "Him gone go bathe" (meaning that he carried out his intention to bathe). The same is true of creoles based on Dutch, French, Portuguese, or Arabic.

Not all linguists agree that creole languages are a promising way to get at a universal grammar—assuming that such a grammar exists. The similarities among them may be due not to an underlying logic but to common

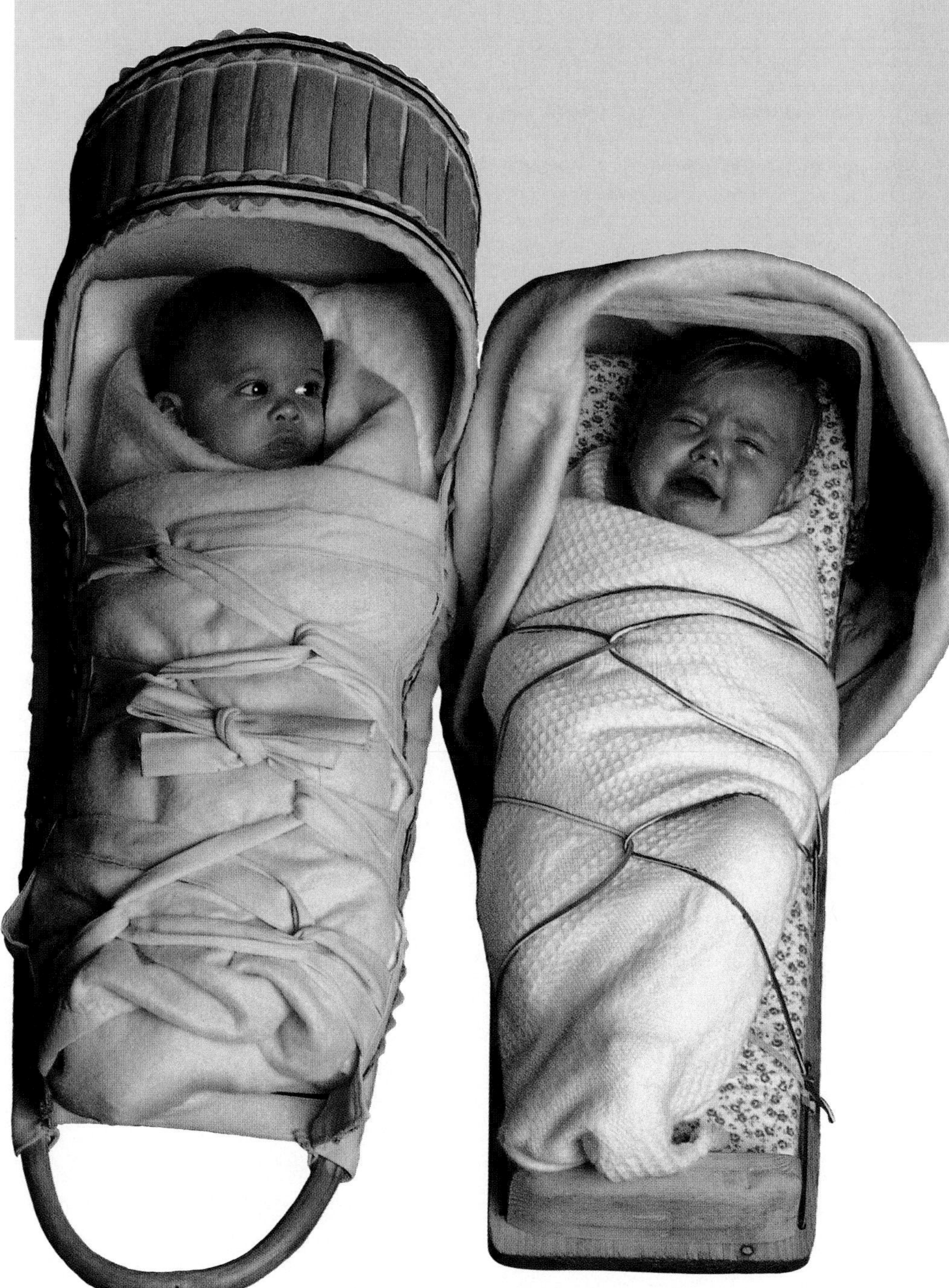

social structures on plantations, they say, or perhaps to the transmission of creole basics from one place to another by sailors. Bickerton concedes that creole tongues may reflect linguistic rules and categories "the next level up" from Chomsky's pure, neurologically imposed grammar. In any case, all theories about a linguistic bioprogram are speculative at this point—a scientific work in progress.

The broader notion that genes are significant determinants of human behavior is more generally accepted. Certainly ample evidence of genetic influence can be found in the animal kingdom—not just pigeons, rats, and other laboratory subjects traditionally favored for behaviorist experiments, but also animals that have been bred for particular traits. Dogs—probably the oldest domesticated species and by far the one most entwined with *Homo sapiens*—tell the story with special clarity. The hundreds of breeds of domesticated dog are so closely related that, with present laboratory techniques, they cannot be distinguished chromosomally from one another (nor, for that matter, from wolves). They nonetheless exhibit great morphological variety—in size, snout length, ear shape, coat color, and so on—and considerable behavioral differences as well. English foxhounds, developed through careful mixing of greyhound, bulldog, and fox terrier stocks, are tireless hunters, whereas Catahoula leopard dogs, a breed perfected along the Gulf Coast of America, express a talent for herding, and Chesapeake Bay retrievers, from the Eastern Seaboard, have a skill for fetching downed game birds.

Refined over generations, the special talents of these hunting, working, and retrieving breeds manifest themselves even when individual dogs are raised away from the hedgerowed fields of Britain, the cattle ranch, or the duck blind. One behavioral sci-

In a study to determine whether ethnic inheritance might include differences in temperament, Dr. Daniel Freedman of the University of Chicago strapped Navajo and Caucasian babies to cradleboards *(left)*. The Navajo babies accepted the new experience calmly, but the Caucasian infants became agitated.

entist, Daniel Freedman of the University of Chicago, reared several sets of pups, including beagles, terriers, Shetland sheepdogs, and African Basenjis, in what he deemed "either an indulged or disciplined fashion" and found that each breed responded differently to a test he engineered. When left alone in a room with a bowl of food and admonished not to eat it, all the Shetlands obeyed, regardless of which way they had been trained, whereas none of the Basenjis did. Among the beagles and terriers, those who had been coddled waited longer to gobble the food than those trained with a firm hand. Freedman posited that temperamental differences among the breeds, deriving from genetic roots, accounted for the results.

From eager-to-please Shetlands and intransigent Basenjis to human children was not a great leap for Freedman. In examinations of infants he found enough tantalizing evidence to surmise that ethnic and cultural differences might have a genetic basis. He discovered that newborn Navajo, Chinese, Japanese, Caucasian, and African babies behaved differently in a number of tests of motor skills and responsivity.

At the time he began his investigations, it was widely thought that all children would display something called the Moro, or startle, reflex, raising their arms and legs and crying if they were lifted in a certain way. It was also assumed that any baby who had a cloth briefly pressed over his nose would struggle to remove it or to free his face, as Caucasian children will. Yet Freedman's subjects consistently failed to support the accepted wisdom. Navajo babies showed little Moro reflex, while Chinese babies who had a cloth placed over their noses simply adjusted to the situation by opening their mouths to breathe. In multiple tests of babies of different ethnicity who were only a few days old, Freedman found other distinctive behaviors that could hardly have been learned.

On the whole, Caucasian babies were the most easily irritated, quickest to cry and slowest to settle down afterward, while Navajo infants showed remarkable placidity and "adaptability" to new stimuli. Chinese newborns fell somewhere in the middle. During test sessions, Freedman reported, "Chinese and Caucasian start to cry at about the same point in the examination, especially when they were undressed, but the Chinese would stop sooner. Furthermore, if picked up and cuddled, Chinese babies would stop crying immediately, as if a light switch had been flipped, whereas the crying of Caucasian babies only gradually subsided."

Motor development also varied among ethnic groups. Lifted onto their feet in a standing position, tiny Caucasian newborns automatically extended their legs and took steps, whereas Navajo infants collapsed their legs. Hoisted into a sitting position by the arms, Japanese and Caucasian newborns did not hold up their heads, but Kikuyu newborns in Kenya as well as newborn Australian Aborigines were able to keep their heads upright. The babies of some ethnic groups simply seemed to develop physical skills more rapidly than others. Another researcher observing Ituri Pygmy children in the African rain forest cited their ability to sit up at three months, walk at six months, and climb trees by one year.

Many of Freedman's findings have been corroborated and amplified by other investigators—often to their discomfort, since suggestions that some cultural differences are genetically based tend to be hotly contested. Freedman himself, recognizing the touchy nature of this argument and how science can inadvertently fuel racial prejudices, cautions against careless logic that might equate genetically based group skills with inborn superiority. Saying that ethnic differences exist is hardly the same as linking them to the overall worldly success or failure of the group, much less the capacity of an individual from that group to function. He observes, for example, that "there is no genetic likelihood either that two such disparate systems as motor precocity and intelligence are controlled by neighboring genes or that the genes of one system play a major role in modifying the other. It is perfectly possible to be both smart and well coordinated." Addressing his critics head-on, he takes note of "our own cultural tendency to split apart inherited and acquired characteristics. Americans tend to eschew the inherited and promote the acquired, in a sort of 'we are exactly what we make of ourselves' optimism. My position on this issue is simple: We are totally biological, totally environmental; the two are as inseparable as are an object and its shadow."

Perhaps the most widely publicized investigations of the possible genetic roots of behavior have been twin studies. For purposes of statistical comparison, such studies typically include both identical twins (technically known as monozygotic, meaning one fertilized egg) and fraternal twins (dizygotic, from two fertilized eggs), and they examine twin pairs that have been raised apart as well as those that have been raised together. Among these different categories of twinhood, one group is especially important for purposes of establishing the relative impact of heredity versus environment: monozygotic twins who have been separated early in life and placed with different families. The presumption is that any behavioral similarities between such separated siblings can be attributed to genetic influences, since the twins were not reared in the same environment.

A well-known arena for twin investigations is the University of Minnesota, where in 1979 psychologist Thomas Bouchard launched the Minnesota

Telling Tales of Twins Rejoined

When the ancient Greeks looked for patterns in the stars, one particular stellar grouping reminded them of the mythical twins Castor and Pollux (*right*); they named the group's two main stars for the brothers and called the entire constellation Gemini, meaning "the Twins." Today, twins themselves are the focus of some intensive pattern seeking. Scientists are using them to assess the relative contributions of genes and the environment to a broad range of human traits, including such behavioral matters as intelligence and personality. One of the better-known efforts in this quest is a long-running research project called the Minnesota Study of Twins Reared Apart, launched in 1979 by psychologist Thomas Bouchard of the University of Minnesota. The prime subjects are identical and fraternal twins who were separated shortly after birth and reared in different adoptive homes.

Bouchard and his team have examined more than a hundred sets of such twins—and a few sets of triplets as well. Although the researchers' objective is to identify broad statistical patterns of heritability rather than to draw conclusions from specific cases, these siblings—some of whose stories are sketched on the following pages—have proved irresistibly fascinating as individuals. Often mirroring one another unawares, the twins have suddenly discovered that, in a sense, they exist in another version.

Study of Twins Reared Apart—a project that continues to this day. At the Minnesota lab, newly recruited subjects undergo a week-long series of physical and psychological assessments. They are fitted with portable devices that keep track of their activity and body temperature around the clock. They are x-rayed, watched with videocameras, placed on treadmills. Trained interviewers record life histories—the family background of the twins, their schooling, sexuality, stressful experiences, and so on. The subjects take standardized tests designed to gauge personality, IQ, and other mental abilities. By the time their stay is over, they have responded to more than 15,000 questions.

In this manner, Bouchard and his team—ophthalmologists, dentists, and cardiologists as well as psychologists and geneticists—have amassed a large database that today includes, as its core, more than a hundred sets of twins raised apart. During interviews of the twin pairs, the researchers have discovered match-ups that seem nothing short of uncanny—shared tastes and inclinations ranging from favorite colors to the names of spouses and children, as in the well-known case of the Jim twins (*page* 8). But as with twin studies done elsewhere, the Minnesota approach is intensively statistical and ultimately concerned with general patterns on scientific measures rather than the idiosyncrasies of specific cases.

Some of the Minnesota results have simply confirmed features of twinship that might have seemed likely to a casual observer. For example, the researchers have not been especially surprised to find that identical twins reared apart tend to be very close in body weight or that they often wear corrective lenses of the same prescription. But certain physical resemblances documented by the Minnesota team have been less obvious.

Identical twins reared apart have very similar immune systems, as measured by the presence of certain antibodies. They show similar susceptibility to heart and lung disease, and they tend to develop illnesses at about the same age. They also have closely matching brain-wave patterns: EEGs of identical twins raised apart are as similar as those of a single individual tested from one year to the next. All such similarities, of course, argue for a strong genetic influence on these traits, since genetic endowment, rather than environment, is what the twin pairs have in common.

Mental abilities, including the notoriously tricky issue of intelligence, have been a prime focus of the Minnesota studies. The Minnesota team has approached the question of the influence of genes on intelligence with a wealth of data: The researchers measure intelligence by a number of widely used tools, including the Wechsler Adult Intelligence Scale and specific tests for verbal, spatial, perceptual, and memory skills.

This information is manipulated by mathematical methods to produce estimates of what behavioral geneticists call heritability, an often misunderstood term. Heritability does not apply to any one individual's genetic inheritance; rather, it refers to the percentage of the variability for a particular trait that may be due to genetic

The Giggle Twins

The lives of Daphne Goodship (*right*) and Barbara Herbert (*left*) parted when their mother, a Finnish student, gave them up for adoption shortly after their birth in London in 1939. Barbara was raised in the family of a gardener; Daphne's adoptive father was a metallurgist. Not until 1980 did they meet again: Barbara had learned from adoption records that she had a sister, and she tracked down her identical twin.

Informally at first, then as subjects for the Minnesota research program, they discovered just how much alike they are. They score almost identically on vocabulary tests, despite the different educational levels of the households in which they were raised. Their physical resemblances extend to a shared heart murmur, slightly enlarged thyroid glands, and matching brain-wave patterns. Certain other parallels in their lives may well be coincidental but are nonetheless uncanny: For example, each twin injured an ankle by falling down stairs at the age of 15, and each met her husband at 16. Each suffered a miscarriage with her first pregnancy, then went on to have two boys and a girl. They like their coffee black, no sugar—and cold. They are both careful with money, and both are afraid of heights. Each twin presses her left hand to her face when she is nervous. Each uses her hands a lot when she talks—and tries to keep them still. The Minnesota researchers call them the Giggle Twins because, as they themselves say, "we laugh more than anybody else we know." Concludes Bouchard, "We see more differences within families than between these two."

A Firehouse Duo

Volunteer fire chief Jerry Levey (*left*) received his first inkling of the existence of a twin brother when he was attending a firefighters' convention in New Jersey in the mid-1980s. Another firefighter insisted that he knew Levey's twin, and he arranged for them to meet. When Mark Newman (*right*) walked into Levey's firehouse, all doubt vanished. "Lop off the extra pounds," Levey recalled, "and I was looking in the mirror. We had the same mustache, the same sideburns, even the same glasses." They stared at each other, let out a simultaneous whoop, and then "partied for three days straight." In that first celebration, they discovered that they both drank only Budweiser beer, both stretching a pinky across the bottom of the can—and crushing the empty.

Separated five days after birth, they grew up in different parts of the New York area. Levey went to college and Newman did not, but in countless other ways they are alike. Both are bachelors and good-time guys, attracted to tall, slender, long-haired women. They like John Wayne movies, Chinese food late at night, and pro wrestling. Both favor the death penalty and oppose gun control. Says Newman, "We agree on 99 percent of things." They even share an avocation: Newman, like Levey, is a volunteer fire captain—perhaps in part because of genes. Some researchers believe that tendencies toward altruistic behavior and risk taking can be inherited; if so, something in their identical genetic endowment may urge the brothers to go to dangerous places to help other people.

differences among individuals in a given population. For example, the Minnesota team estimates that heritability for IQ is about 70 percent, meaning that 70 percent of individual variation may be attributed to genetic, rather than environmental, factors.

As it happens, the Minnesota findings are higher than those of a number of other twin studies, which find the heritability of intelligence to be about 50 percent, on average. On tests of other mental abilities, the Minnesota researchers found an average heritability of about 50 percent, with spatial abilities the highest and visual memory the lowest.

If intelligence displays a fair degree of genetic influence, what about personality? Bouchard's team was not the first to examine personality traits by means of twin studies. In 1976 researchers at the University of Texas performed a personality assessment of 514 pairs of identical twins and 336 pairs of fraternal twins who had taken the National Merit Scholarship qualifying test some years earlier and arrived at an average heritability estimate of 50 percent for the personality traits revealed by this data.

The Minnesota researchers have accumulated personality data through standardized tests that assess a subject's overall sense of well-being, aggressiveness, ability to deal with stress, degree of extroversion or introversion, acceptance of traditional values, amount of self-control, psychological flexibility, and many other traits. Their results conform to those of the University of Texas researchers. Summing up the findings, Bouchard says that the personality resemblance of identical twins "does not appear to depend on whether the twins are reared together or apart, whether they are adolescent or adult, [or] in what industrialized country they reside." He noted further that "approximately 50 percent of the variance in self-reported personality characteristics is associated with genetic factors, which means that environmental factors account for an equal share of the variance." Fraternal twins reared together offer supporting evidence: Their median personality correlation was half that of identical twins—exactly what is expected from subjects that share about 50 percent of their genes rather than 100 percent.

Contrary to conventional wisdom, studies of both identical and fraternal twins suggest that the environment achieves its influence in ways that are unique to each individual, even within the same family. The personalities of children reared together are formed far less by what they share in their environment (a common socioeconomic background, schools, home life, and so on) than by their own particular interchanges with parents and

A Deep Divide

Of all the identical twins studied at the University of Minnesota, Jack Yufe (*right*) and Oskar Stohr (*left*) described the most disparate childhoods. Born in 1933 in Trinidad to a Jewish father and a German Gentile mother, the boys were soon separated by their parents' divorce. Jack was raised by his father as a Jew in the Caribbean and Israel. Oskar was brought up Catholic by his grandmother in Hitler's Germany. The two knew of each other but met only once, in 1954; the fact that Jack had just come from a kibbutz, while Oskar denied his Jewish heritage, made this first meeting a tense one.

Jack settled in California, eventually owning a clothing store; Oskar became a factory supervisor in Germany. Their wives kept in touch with occasional cards. Then Jack learned of Bouchard's project; with Bouchard's help, Oskar was persuaded to participate. In Minnesota for a week's testing, the men had a happier reunion. They arrived sporting identical eyeglasses, short mustaches, and two-pocketed shirts with epaulets, and went on to find much else in common. Both liked spicy foods and sweet liqueurs. Each has the unusual habit of flushing the toilet before as well as after using it; each finds it funny to startle strangers with a loud, fake sneeze. Both nod off to sleep in front of the television, read magazines from back to front, and store loose rubber bands around one wrist. But, not surprisingly, in some of their attitudes they are very different. Jack found Oskar "very domineering" toward women; Oskar found Jack "too soft and warm-hearted" to his employees.

Alike but Not

April Yamashiro (*left*) and Linda Benton (*right*) presented a mixed picture for Bouchard's team of researchers. Born in Japan and raised by separate families in California, they seemed a matching pair in many ways. For example, both have had miscarriages, both suffered from a similar intestinal ailment, and both have a split nail on the same toe.

Yet Yamashiro has a visual defect that requires her to wear glasses, and Benton does not. Only one twin is afraid of flying. One is outgoing, the other rather reserved. At first, noting dissimilarities in the shapes of their faces and ears, Bouchard's team wondered whether the women were in fact genetically identical. A blood test established that they are.

Yamashiro and Benton exemplify the complexity of the interplay between genetic and environmental factors in human development. The interaction begins immediately after conception and continues throughout life, and it may affect even highly heritable traits such as facial shape. Twins may encounter particular environmental stress in the womb, precisely because they are twins. The two developing fetuses compete for nutrients, and one may get a greater share of the blood supply. In the case of Benton and Yamashiro, the Minnesota researchers were less concerned with how the differences arose than with the nature and degree of those differences—a caution against seeing genes as predestination.

peers and their own individual set of experiences. As Minnesota psychologist Auke Tellegen says: "Our findings don't show that families have no influence over their children. Families undoubtedly affect children, but not uniformly." Sandra Scarr, a developmental psychologist now at the University of Virginia, puts it this way: "Upper-middle-class brothers who attend the same school and whose parents take them to the same plays, sporting events, music lessons, and therapists and use similar child-rearing practices on them are little more similar in personality measures than they are to working-class or farm boys, whose lives are totally different." In fact, a shared environment appears to have a differentiating effect on identical twins: A number of studies have shown that identical twins raised together are less alike in personality than those raised apart—presumably a consequence of their efforts to forge a distinct identity.

Considering the evidence for genetic influence on intelligence and personality, it is not surprising that numerous twin studies indicate a link between genes and mental health, from relatively minor disorders such as phobias and mild depression to severe illnesses such as schizophrenia and manic depression. At Minnesota, the researchers have found that identical twins reared apart have very similar psychiatric histories—eerily parallel in some cases. For instance, one pair of twins revealed a shared fear of water—and a shared tactic for dealing with their phobia: Both of them always backed into a lake or pool.

Numerous investigations in the mental-health arena have concentrated on schizophrenia, a disease that strikes about one out of every hundred Americans, most of them between the ages of 18 and 24, and that afflicts about 10 million people worldwide. In the majority of cases, the symptoms are devastating and lifelong. Many patients hear voices, which may utter a stream of threats or issue dire commands or perhaps seem to emanate from God. Hallucinations are common, and some patients become profoundly withdrawn.

The origins of the illness have long been in dispute. Some theorists argue for a physical cause—some organic defect of the brain—and others emphasize such environmental factors as harsh, overprotective, or rejecting parents. There has never been any doubt that schizophrenia runs in families: One out of 10 children with one schizophrenic parent develops the illness, and the rate rises to almost one out of two when both parents are schizophrenic. These figures, of course, could be cited as support for either the nature or nurture position,

A Three-Way Tie

Nineteen-year-old Bobby Shafran (*left*) found his first days at Sullivan County Community College in upstate New York thoroughly confusing. Other students greeted him like an old friend—but they called him "Eddy." The reason became clear when he was shown a picture of someone who had attended Sullivan the year before, Eddy Galland (*right*). Bobby could have been looking at himself.

The brothers had been adopted as infants and met as total strangers, yet they seemed to have lived copycat lives. Each had flunked math in fifth grade, had taken up wrestling, had gotten psychiatric help, and had been told his trouble stemmed from being adopted. The story was remarkable enough to make the newspapers—and then it became unique.

Yet another 19-year-old college student, David Kellman (*center*) at Queens College, saw the story and picture and perceived himself in it. The young men made medical history as the first identical triplets to be separated soon after birth and reunited. David, too, had flunked fifth-grade math, enjoyed wrestling, and received psychiatric help. Though all grew up in Jewish homes, they all favored Italian food. All smoked heavily and dated older women. They talked alike, laughed alike, and had matching birthmarks and IQs of 148. Instantly famous, the three enjoyed a round of television appearances. But for at least one of them, the new life was a mixed blessing. "All my life I have felt special and individual," said a wistful Bobby Shafran. "Now I've met two people just like me."

Questions of Uniqueness

Barbara Parker (*right*), a landscape designer, and Ann Blandin (*left*), a homemaker, discovered one another in Los Angeles in 1983, when Parker's neighbor introduced them. For both the 36-year-old women, recognition of their twinship was instantaneous. Their cheekbones, their smiles, their gestures, and their laughs were alike.

In personality and attitudes, however, the match-ups proved to be only partial. Although both women are sociable, emotional perfectionists, Blandin considers herself more critical than her twin and more vulnerable to criticism; she also has stronger religious feelings. Not surprisingly, their adoptive childhood environments were quite different. Blandin recalls feeling like the "oddball" skinny brunette, left behind as her three tall, blond sisters grew up glamorous and career minded. By contrast, Parker was raised as a beloved only child, first by her adoptive mother, who died when Parker was nine, then by her longtime baby-sitter.

For Dr. Thomas Bouchard, most often the point is not how much separated twins diverge but how little. "Differing experiences have made them unique and different," he says, "while identical genes have made them uniquely alike in spite of those differences."

but some other statistics—drawn from a series of studies conducted in Europe over a period of decades—have made a compelling case for genetic influence. They show that, on average, relatives of schizophrenics are at risk of developing schizophrenia themselves in proportion to their degree of genetic relatedness.

Twin studies appear to underscore the point, indicating that the odds of the identical twin of a schizophrenic also developing schizophrenia may be anywhere from 40 to 90 percent. The degree of risk appears to be much greater than that faced by a fraternal twin of a schizophrenic, and it applies with full force to identical twins reared apart as well as to those raised together. Still, psychiatrists are by no means certain how genetic influences are expressed to produce schizophrenia. Most likely, a number of genes are involved, and they may only establish a predisposition to the illness. According to Irving Gottesman, a professor of psychology at the University of Virginia and a leading investigator in the field, "The most likely explanation is that the right combination of genes—probably four or five—plus some as yet undefined environmental stressors must be thrown together to trigger schizophrenia."

Alcoholism is another common disorder that may have some genetic basis, although the evidence is even less clear than for schizophrenia. In the 1970s, studies showed that children of alcoholics who were adopted by nonalcoholics were more likely to become alcoholics themselves than their nonbiological siblings. The effect was especially pronounced with sons. Several recent studies have focused on females, with dramatically different results. One analysis of identical twins raised together indicated that a woman whose identical twin sister became an alcoholic was five times more likely to be an alcoholic than the average person; with fraternal twins, as would be expected, the odds dropped to 1.6. With female identical twins raised apart, however, a University of Minnesota study found that, unlike the case for males, the estimated heritability for alcoholism was zero.

Similar uncertainty surrounds the biochemical basis of alcoholism. In 1990 autopsies carried out on the bodies of deceased alcoholics by scientists at the University of California, Los Angeles, and at the University of Texas produced what seemed to be a promising lead. The brain cells of 69 percent of the cadavers possessed a special variant of a gene that codes for the so-called D2 receptor, a molecular docking site for the neurotransmitter dopamine; among nonalcoholics, the incidence of the variant gene for the D2 receptor is only 20 percent. News of this finding spread rapidly, and excited media reports ensued, proclaiming that the biochemical culprit in alcoholism had been nailed down. However, additional studies have challenged this conclusion. A number of researchers see no link at all between the D2 gene and alcoholism.

Genetic explanations have also been offered for criminality. After a large-scale survey in 1961 reported that men with an extra Y chromosome were overrepresented in prison populations as compared with the male population at large, some criminologists hypothesized that the extra Y chromosome, conveyed to a male at conception, made him, in effect, a hyperaggressive "supermale" and predisposed him to a life of violence. Numerous follow-up studies failed to support the connection, however, and the idea was dismissed by most scientists by the early 1970s. Still, some researchers continue to nurse the hope of identifying a genetic marker that might allow investigators to spot those with a predisposition for traits that might be linked to criminality.

Not everyone believes that the search for such a marker would be conducted in a disinterested manner, even if it had any possibility of success. In September 1992, a conference funded by the National Institutes of Health to explore the genetic roots of criminality was summarily canceled by the agency because of its perceived racial implications. A spokesperson for the NAACP, which had joined several other groups in urging the agency's action, explained that the organization feared that the conference supported efforts to "relate crime to the African-American community, and to say there are ingrained or genetic reasons why we are more prone to crime than others." Although investigations continue into the genetics of violent crime, psychologist Glenn D. Walters of the Federal Correctional Institution-Schuylkill in Minersville, Pennsylvania, has suggested, after reviewing 38 studies carried out since the 1930s, that the results do not justify the time and money spent in the pursuit. "I don't think we will find any biological markers for crime," he declared in the November 1992 issue of the journal *Criminology*. "We should put our resources elsewhere."

Few fields of science have generated more controversy than behavioral genetics, but most researchers accept that it comes with the territory. The notion that heredity may be critical in influencing the development of an individual has stirred heated debates ever since the days of Sir Francis Galton, and efforts to assess the relative influences of nature and nurture in such complex and sensitive spheres as intelligence, personality, or mental health have hardly calmed the arguments. Charges of badly designed experiments, flawed tests, poor handling of data, and fundamental logical errors are commonplace.

Studies of identical twins raised apart—the source of some of the most intriguing results in behavioral genetics—have not been exempt from attack. Critics point out that, although separated shortly after birth or during the first years of life, few such twins have been reared in wildly disparate cultures. More often than not, they have been placed with relatives or matched with adoptive families having concordant economic standing and values. Given such circumstances, say critics, it is all but impossible to cleave genetic influences from environmental ones.

Richard Lewontin, a population geneticist at Harvard University, emphasizes that "the organism is the outcome of a process of development." The process begins before birth: "Of the physical and physiological differences among individual people that are present from birth, many are caused not by genetic differences but by developmental noise"—that is, by environmental factors that can affect a growing fetus. He gives the example of fruit flies, which hatch with eyes of different size depending entirely on the temperature they have been exposed to during their immature phase. The human fetus seems particularly sensitive to damage from some avenues during the first trimester, and women who drink alcohol during this phase of the pregnancy may harm the fetus permanently, giving birth to babies with physical and psychological defects.

The developmental process never stops. "The individual living being is, then, at every moment the consequence of a unique interaction between its genotype and the history of the environments in which it has found itself," Lewontin says. Individuals do not necessarily "find themselves" in particular environments by happenstance, of course. In regard to personality, for example, "all of us make our own environment," says psychologist Sandra Scarr. Or as David Lykken of the University of Minnesota puts it, "The environment molds your personality, but your genes determine what kind of environment you have, seek and attend to."

For society in general, observes

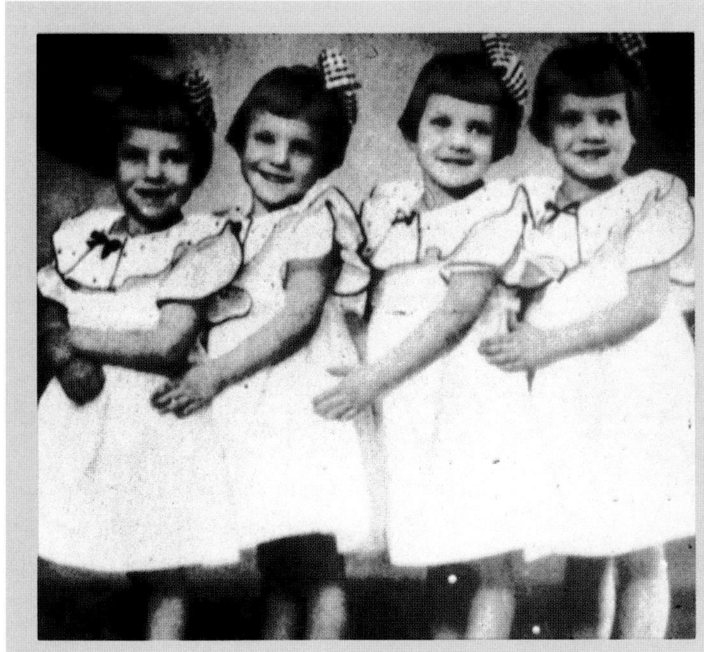

All four of these identical quadruplets, born in the American Midwest in the 1930s and raised in a troubled household, suffer from schizophrenic delusions. However, their symptoms differ—suggesting that both environmental and genetic factors play important roles in the disease.

Lewontin, a particular danger lies in the tendency of nonspecialists to equate heritability and immutability. Numerous experiments have demonstrated that, while IQ seems to be about 50 to 70 percent heritable, children who receive special tutoring can boost their scores significantly. Similarly, to say that alcoholism is heritable does not mean that every son of a besotted father is doomed to repeat the destructive familial pattern. Too often, notes Lewontin, heritability is assumed to mean insensitivity to environmental change. The geneticist contends that "the problem goes back to the false dichotomy between nature and nurture, to the belief that gene and environment are separate and separable determinants of organism rather than interacting and inseparable shapers of development."

Most geneticists are careful to point out that genes—where they appear strongly linked to behavior—are not mandates but flexible programs that code for a possible range of outcomes. "The basic concept for a correct understanding of gene and organism," says Lewontin, "is the norm of reaction. For a given genotype, there will be a particular phenotype for each environment. The norm of reaction for a genotype is a list or graph of the correspondence between the different possible environments and the phenotypes that would result." To take a physical example, the genotype may set limits on height—a maximum and a minimum depending on diet; if food is ample, a person may reach the top of the range, whereas if food is scarce he may never grow far past the lower mark. The same is true for behavioral traits that are influenced by both genetic and environmental factors.

The degree of responsiveness among genotypes, as expressed in the phenotype, differs. For some people, slight shifts in one or more external factors translate into large phenotypic changes. Obesity may be an example. According to one school of thought that draws on data from twin studies, genes, environment, and a sedentary lifestyle combine to make people overweight. Identical twins who served as subjects in an experiment carried out in Quebec in the late 1980s were put on the same diet and restricted to the same low level of activity. The members of each twin pair uniformly gained equal amounts of weight, putting on the pounds in pretty much the same places. Between the sets of twins, though, there were wide divergences in pounds added. Eating the same high-calorie diet, a set at one extreme ballooned 30 pounds; at the other extreme, another set gained only nine. In an industrial society offering an abundance of food and every excuse to avoid exercise, people who are, in effect, genetically at risk to gain excess weight must work harder not to do so.

In other cases, environmental inputs may provoke only tiny phenotypic alterations, yielding minor divergences in height, weight, immune function, or other characteristics. For example, Caucasian infants are typically more boisterous and vocal than Japanese ones. This seems partly due to an innate quietude on the part of Japanese babies, which is further reinforced by the relative docility of most Japanese mothers. When Japanese migrate to the United States, they often begin to adopt the noisier habits of their new countrymen, and mothers become correspondingly more talkative with their infants, prompting the babies themselves to sound off more. However, studies have revealed that to prompt Japanese infants to the same level of vocalization as that routinely engaged in by Caucasian infants requires Japanese mothers to work twice as hard at stimulating their children as Caucasian mothers. In other words, it is harder to nudge Japanese babies into greater vocal responsiveness.

Daniel Freedman argues that, in light of the facts, "only a truly holistic, multidisciplinary approach" to behavioral science makes sense. Humans, he says, are "biosocial" creatures, shaped, as he contends, 100 percent by biology, 100 percent by culture. So-called interactionist theories navigate between the horns of the nature-nurture dilemma by recognizing that, in the words of one personality expert, "the contributions of heredity and the environment are not merely additive but rather complex combinations." Every infant is born with proclivities and aversions and may be moody or stoical, introverted or extroverted. One child may routinely respond happily and vigorously when loud music is played in the house; another may scowl with annoyance and burst into tears. These aspects of personality do not express themselves in a vacuum but instead may become apparent depending on the manner in which the infant is treated. A depressed mother, whose sadness, irritability, and lethargy may cause her to punish or neglect her infant, can over time have a marked effect on her baby's entire personality. Repeatedly unable to get his needs met through his mother, the child may himself withdraw into depression out of a sense of powerlessness to interact positively with the world.

Given the influence of genes, it is probably the case that a parent cannot utterly alter a child's predilections, but instead only subtly reinforce or downplay traits. Shy children, for example, shrink from new stimuli, clinging to their parents in the face of strangers or unfamiliar situations. However, a tendency toward shyness, which, studies suggest, may carry a strong genetic component, does not fate children to be retiring wallflowers as adults. Instead, says Harvard psychologist Jerome Kagan, children can be helped to overcome their discomfort, although they should not be made to feel self-conscious. Kagan advises, "Parents need to push their children—gently and not too much—into doing the things they fear."

Does the hardware of the brain, sketched by the genetic program, carry specifications for such complex human traits as religiosity or particular political attitudes? How will scientists sort out which traits are owed to the operation of one or two genes, which to dozens? The truth is that no one knows how far geneticists will be able to go in illuminating human behavior. In just over 100 years, researchers have progressed from abject ignorance of the workings of heredity to a degree of knowledge that no doubt would have impressed Sir Francis Galton. Innumerable superstitions and hoary folk beliefs have been dismissed in the process, yet misbegotten theories concerning inbred genetic superiority of some groups persist, debunked but not yet fully quashed. Lewontin, for one, warns against misguided social programs, based on the erroneous notion that social strata and racial divisions are somehow built into the genes. Such an argument, he says, "is essentially an argument for the inevitability and justice of the status quo. It is fairly obvious who the argument serves."

What remains for science is to puzzle out the chain of relationships between genotypes and phenotypes: to isolate genes, identify their biochemical products, show how those products act to modify bodily structures or processes, and, finally, reveal how a brain system or an excess of a certain neurotransmitter regulates or alters behavior under certain environmental conditions. Most likely, future gains in knowledge about the intricacies of human nature will not diminish but rather will enhance respect for the individual. As Bouchard remarks about his much-perused twins, "No matter how similar we found them to be, their fundamental uniqueness has always shown through."

Unraveling the Secrets of Life

People have long been fascinated by the vagaries of heredity, the mysterious way some traits pass unfailingly from one generation to the next, while others skip a generation or two—or appear seemingly out of nowhere. In the latter half of the 19th century, as a result of the work of Gregor Mendel, fascination turned to science. Using only native intelligence, painstaking attention to detail, and some simple math, the Austrian monk discovered the laws of inheritance by careful analysis of many generations of pea plants. Without being able to observe them directly, Mendel postulated the existence of physical factors—now known as genes—that carry the blueprint for all manner of traits from parents to offspring. Half a century later, biologists would begin the difficult process of homing in on this blueprint and discovering its language in the complex molecule known as DNA.

The breakthrough discovery of DNA's chemical structure in 1953 not only laid the groundwork for a better understanding of how heredity works, but also suggested the potential—realized a few decades later—to alter genes themselves. These achievements have paved the way for the Human Genome Project, an intensive effort to map the entire sequence of nucleotides that make up human DNA and determine the function of every gene. By locating where on the chromosome a given gene lies, and learning how to modify any faults in its makeup, geneticists hope eventually to find remedies for genetic disorders and cures for now-fatal diseases.

1859

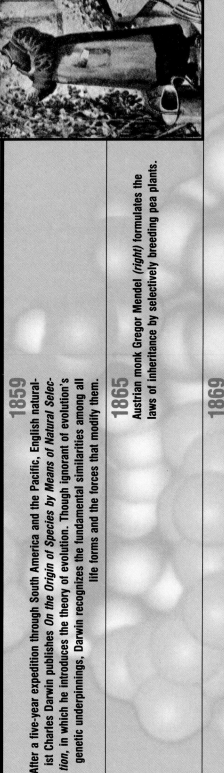

After a five-year expedition through South America and the Pacific, English naturalist Charles Darwin publishes *On the Origin of Species by Means of Natural Selection,* in which he introduces the theory of evolution. Though ignorant of evolution's genetic underpinnings, Darwin recognizes the fundamental similarities among all life forms and the forces that modify them.

1865

Austrian monk Gregor Mendel *(right)* formulates the laws of inheritance by selectively breeding pea plants.

1869

Swiss physiologist Johann Friedrich Miescher inadvertently isolates DNA from the nuclei of white blood cells. Although Miescher and others carefully study the "nuclein," its function eludes them.

1873

Improved microscopes and staining techniques let scientists view unusual thread-like structures within cell nuclei. The behavior of these structures, dubbed chromosomes for their affinity for colored stains, hints at some role in cell division.

1883

Francis Galton, a cousin of Darwin's, proposes that intelligence is inherited just as physical features are. He champions a practice he calls eugenics—the selective breeding of humans "to give the more suitable races or strains of blood a better chance of prevailing speedily over the less suitable."

1903

American graduate student Walter Sutton and German biologist Theodor Boveri independently propose that chromosomes in the cell nucleus *(right)* contain the vehicles of inheritance.

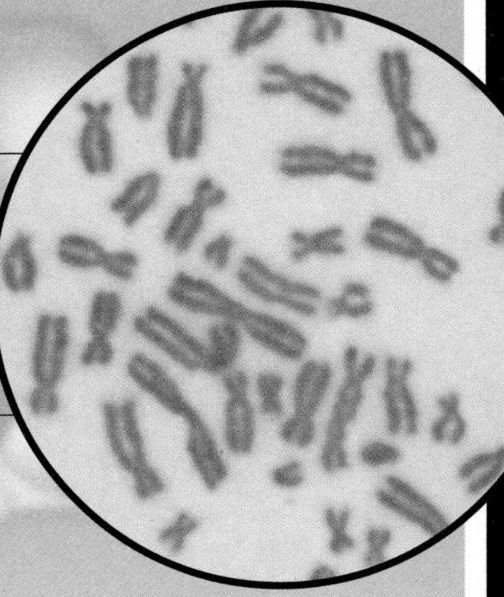

1909

The term *gene,* used to describe the fundamental unit of heredity, is introduced by Danish biologist Wilhelm Johannsen in his book *The Elements of Heredity.*

1910

While experimenting on fruit flies in his Columbia University laboratory, Thomas Morgan identifies an eye-color mutation that passes from parent to offspring through the X sex chromosome. He becomes the first experimenter to link a specific gene to a specific chromosome.

1944

Canadian-born physician Oswald Avery *(left),* while studying the bacterium that causes pneumonia, proves that genes are made up of DNA.

1950

Columbia University biochemist Erwin Chargaff, studying the chemical composition of DNA, uncovers an unexpected relationship among its four nucleotides: The molecule always contains equal quantities of cytosine and guanine, and equal quantities of adenine and thymine. Having little interest in genetics, however, Chargaff leaves it to others to explore the implications of this finding.

1953

Building on the work of Chargaff and others, James Watson and Francis Crick *(right)* decipher the double-helix structure of DNA. They correctly propose that genetic information lies encoded in the arrangement of nucleotides in the molecule.

1956
At the University of Lund in Sweden, researchers Joe-Hin Tjio and Albert Levan show that human cells contain 46 chromosomes—rather than 48, as previously thought—arranged in 23 pairs.

1959
In the first example of tracing a disease to a certain genetic abnormality, French scientist Jerome Lejeune discovers that Down syndrome is caused by the presence of an extra chromosome 21.

1961
At the United States National Institutes of Health, Marshall Nirenberg, Johann Mattaei, and Philip Leder crack the "genetic code"—the specific nucleotide sequences within DNA that code for each of 20 amino acids essential to human life.

1968
British biologist John Gurdon clones frogs by replacing the nucleus of a frog egg cell with the nucleus of a cell from another frog's tadpole; the egg cell and tadpole grow into identical-twin frogs. Later, multiple clones *(left)* are produced from skin cells.

1973
Geneticists splice together DNA segments from a toad cell and a bacterial cell and implant the hybrid in another bacterial cell, enabling mass production of the toad's DNA and its proteins. Their success gives rise to the science of genetic engineering.

1975
At the Asilomar conference in California, 150 of the world's leading geneticists discuss the potential hazards of their work and draft guidelines to reduce the risk of public harm from genetically engineered organisms.

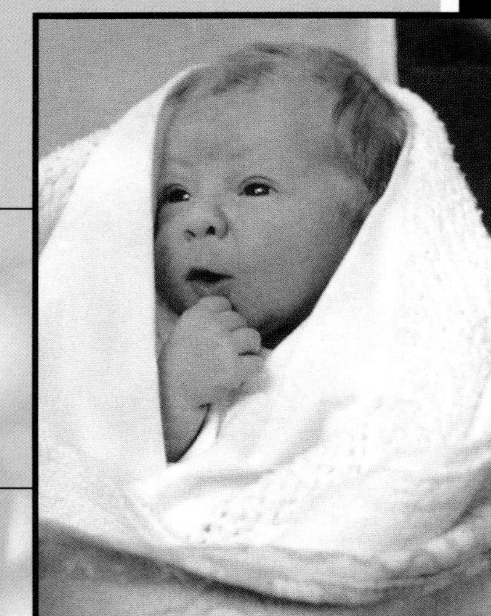

1978
Louise Brown *(right),* the world's first "test-tube baby," is born in England. Because Brown's mother could not conceive a child naturally, doctors removed an egg from one of her ovaries, fertilized it in a laboratory with sperm from her husband, and implanted the embryo in her uterus.

1980
In a ruling of great importance for the fledgling biotechnology industry, the United States Supreme Court rules that genetically engineered life forms can be patented.

1982

Biologists genetically engineer mouse eggs to produce high levels of growth hormone. The result is a strain of "supermouse" *(far left)*, which can grow nearly twice as large as normal mice *(left)*.

A procedure known as chorionic villus sampling is first used to test for genetic defects in unborn babies. The procedure involves the removal and analysis of a sample of tissue that will eventually develop into the placenta; any anomalies in the tissue's DNA will also be present in the fetus.

1984

Researchers at the University of Indiana and Switzerland's University of Basel discover that a similar DNA sequence occurs in a number of growth-controlling genes in the fruit fly. The sequence, named the homeobox, has counterparts in the genes of other species—a finding that sheds new light on how animals develop.

The innovation of genetic fingerprinting allows forensic scientists to compare DNA from the blood, hair, or semen left at a crime scene with that of a suspect. Three years later, prosecutors win their first conviction with this technique.

1990

In the first attempt at human gene therapy, doctors at the National Institutes of Health transfuse a young girl with white blood cells *(right)* genetically engineered to manufacture an essential immune protein her own cells do not produce.

1992

Calgene, an American bioengineering firm, produces a genetically engineered tomato—one that tastes better and has a longer shelf life than ordinary tomatoes.

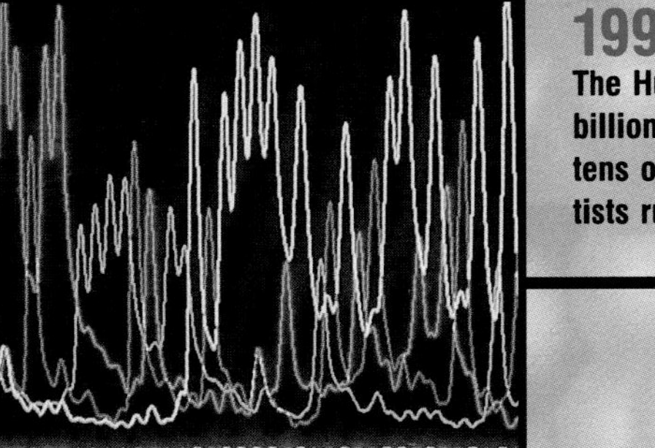

1993

The Human Genome Project, begun in 1990, sequences 120 million of the three billion nucleotide pairs in the human genome. With computers that can sequence tens of thousands of bases each day (a sample readout is shown at left), the scientists running the project hope to finish sometime in the 21st century.

GLOSSARY

Allele: one of two or more different versions of a single gene; from a Greek word meaning "other." The gene that determines blood type, for example, has three alleles: A, B, and O. A person normally possesses two alleles of each gene, one from each parent.
Amino acid: a class of chemical compounds that are the building blocks of proteins; various combinations of 20 basic amino acids make up all the proteins in the human body.
Antibodies: proteins produced by certain cells of the immune system whose primary job is to neutralize invading pathogens such as bacteria and viruses. Genes are known to code for at least five classes of antibodies.
Autoradiograph: an image produced by exposing x-ray film to material, such as DNA fragments, that has been radioactively tagged.
Autosome: a term used to describe 22 of the 23 pairs of human chromosomes; it excludes the sex chromosomes; from two Greek words meaning "self" and "body."
Bacteriophage: a type of virus that only attacks bacterial cells.
Base: one of the five compounds—adenine, guanine, cytosine, thymine, and uracil—that form the genetic code in DNA and RNA.
Behaviorism: a theory holding that a person's environment plays a more significant role in shaping personality than do genetic factors.
Bystander effect: a phenomenon in which toxins released by a diseased cell as it dies are thought to kill neighboring diseased cells.
Cancer: a disease characterized by the uncontrolled proliferation of cells; it has been linked to mutations in the cells' DNA.
Cell: the basic structural unit of virtually all living creatures. In humans, all cells except red blood cells include a nucleus, which houses the chromosomes that carry the genetic code.
Centrioles: structures within cells that play a crucial role in cell division, generating fibers that pull chromosome strands or pairs of chromosomes apart.
Centromere: the point at which a chromosome's two copies—formed prior to cell division—are joined.
Chromatin: the name given to the loose, ropelike strands of DNA in the nuclei of cells that are not dividing. As a cell prepares to divide, chromatin strands become more compact, in which form they are called chromosomes.
Chromosome: a structure within a cell nucleus that consists of tightly packed DNA. In humans, 23 pairs of chromosomes carry the entire genetic code; each pair consists of one chromosome inherited from each parent.
Codominance: in genetics, a situation in which neither of two traits arising from different alleles of a given gene is dominant over the other. AB blood type is an example of codominance: Both the A and the B allele are expressed, leading to the presence of both A and B protein compounds in blood cells.
Codon: a sequence of three bases along a strand of messenger RNA that represents a specific amino acid.
Courier proteins: proteins that influence cell differentiation by ferrying biochemical signals between developing cells; they are created by homeotic genes.
Crossing over: a phenomenon that takes place during meiosis, in which members of a pair of chromosomes exchange short segments of DNA.
Cytoplasm: the area between a cell's outer membrane and its nucleus, which is filled with a thick liquid that contains a variety of components, including ribosomes and mitochondria.
Developmental noise: any external factors—such as the mother's diet—that can affect the growth of a fetus.
Differentiation: the process whereby dividing embryonic cells begin to produce specific proteins that define the cells' role in the body.
DNA (deoxyribonucleic acid): the complex molecule that, along with structural proteins, makes up chromosomes. It consists of two twisted strands, made up of sugars and phosphates, to which are attached four types of bases—adenine, guanine, thymine, and cytosine; chemical bonds between pairs of bases link the two strands. The sequence of bases along a strand encodes genetic information.
DNA helicase: an enzyme that plays a role in DNA replication, separating double strands of DNA into two single strands by breaking the bonds linking base pairs.
DNA polymerase: an enzyme that completes the process of DNA replication by attaching loose nucleotides to the separated single strands of a DNA molecule, thereby creating two new double-stranded molecules identical to the original one.
DNA profiling: a technique that identifies repetitive sequences in samples of DNA coding, used for distinguishing between individuals or for establishing hereditary connections; also known as DNA fingerprinting.
Dominant trait: a characteristic stemming from an allele that is preferentially expressed over any other alleles for a given gene. For example, the trait for curly hair is dominant over the trait for straight hair.
Enzyme: a type of protein that serves as a catalyst for a specific function, such as digestion or, within a cell, DNA replication.
Gel electrophoresis: a laboratory tech-

nique in which DNA fragments, which carry a negative charge, are placed at one end of a bed of gelatin and are attracted to a positive charge imposed at the other end; because smaller fragments travel faster than larger ones, the DNA pieces sort out by size.

Gene: a length of DNA that codes for the production of a specific protein; in humans, genes typically consist of thousands of base pairs.

Gene expression: the process in which a cell produces the protein encoded by a particular gene.

Gene therapy: a method of treating disease that hinges on the insertion of new genetic material into a patient's cells, affecting cellular function in one of several different ways.

Genetic probe: a molecule consisting of a specific sequence of bases that enables it to attach to a complementary sequence on a DNA strand. Probes typically include a radioactive isotope or fluorescent dye that reveals the sequence's location on an individual strand.

Genetics: the study of heredity, including how traits pass from parents to children and the molecular foundations of those traits.

Genome: the entire set of genetic instructions for a given organism. The human genome, for example, consists of some 100,000 individual genes.

Germ cell: a special type of cell that develops into reproductive cells—sperm cells in males and egg cells in females. In humans, germ cells begin with a full complement of 46 chromosomes, but in one stage of division, called meiosis, that number is halved, so that the resulting reproductive cells contain only 23 chromosomes each.

Heritability: an estimate of the degree to which genetic factors contribute to the manifestation of a particular trait within a population.

Heterozygous: referring to a condition in which the two copies of a particular gene in an organism code for different traits, such as clear skin and freckled skin; in most cases, one of these traits will be dominant over the other.

Homeobox: a short nucleotide sequence common to all the homeotic genes in a given species and identifying them as master regulators of development.

Homeotic genes: genes that regulate the differentiation of cells in a developing organism as well as the relative positions of cells, mapping out the body's overall architecture.

Homozygous: referring to a condition in which the two copies of a particular gene in an organism are the same.

Intron: a segment of DNA along a chromosome that falls between coding regions and that does not itself represent genetic coding; the sequence of bases in introns differs from individual to individual, but their purpose has not yet been determined.

Jumping gene: a gene that migrates from one place on a chromosome to another, or from one chromosome to another, often affecting the function of neighboring genes.

Karyotype: a standard way of organizing the full complement of chromosomes in a cell; a normal human karyotype shows 46 chromosomes arranged in pairs by size.

Linkage map: a type of genetic map that relies on observed associations between genes—often revealed by the inheritance of traits in families—to determine the relative positions of genes on a given chromosome.

Meiosis: a special type of cell division, in which a germ cell splits into two new cells, each containing only half the original number of chromosomes; these new cells go on to develop into sex cells.

Messenger RNA (mRNA): a type of RNA that relays the genetic information encoded in DNA to a cell's ribosomes, which use the information to manufacture proteins.

Mitochondria: structures in the cytoplasm of cells that generate energy to power the cell's activities.

Mitochondrial DNA: DNA residing within a cell's mitochondria that codes for some of the proteins that aid in cellular energy production; in humans and other animals, it is the only form of DNA not located on the chromosomes. Because the mitochondria of sperm cells are in the tail and are not incorporated in the creation of an embryo, offspring inherit only their mother's mitochondrial DNA.

Mitosis: the typical type of cell division, in which a cell splits into two exact copies, each containing a full complement of chromosomes.

Mutation: any alteration in an organism's genetic code, from one base in a gene's sequence to the number of chromosomes in each cell.

Nucleotide: the fundamental unit of DNA and RNA, consisting of a sugar-and-phosphate compound joined to one of the five bases.

Nucleus: a specialized component of almost all cells that contains, among other structures, the cell's chromosomes; it is encased in a membrane and often is located at or near the cell's center.

Oncogene: a gene that normally regulates cell growth and development but, if mutated or improperly activated, can give rise to cancer.

Protein: a molecule consisting of up to thousands of amino acids linked together and folded to form a distinct shape that determines the protein's function. Proteins are the fundamental components of the body and play an essential role in all biological processes.

Recessive trait: any trait that depends for its expression on the presence of two identical alleles of a given gene. A recessive trait will not manifest itself when its allele is paired with the allele responsible for the dominant version of that trait.

Recombinant DNA: a DNA molecule synthesized by combining a segment of DNA from one source with DNA from another source. In gene therapy, recombinant DNA technology involves implanting a therapeutic gene into the DNA of a microorganism so that it will be mass-produced.

Regulatory gene: any gene that codes for a protein involved in regulating the function of other genes.

Replication: the process whereby DNA makes an exact copy of itself.

Restriction enzyme: an enzyme that cuts a strand of DNA wherever the enzyme recognizes a specific sequence of bases.

Retrovirus: a type of virus, such as the one responsible for AIDS, that contains RNA rather than DNA. Unlike DNA-containing viruses, retroviruses must first convert their RNA into DNA before their genetic instructions can be carried out.

Ribosome: a structure within cells that manufactures proteins by linking together amino acids according to the coded sequence on a strand of messenger RNA.

RNA (ribonucleic acid): a complex molecule closely related to DNA; RNA consists of a single rather than a double strand, is composed of a different type of sugar, and substitutes the base uracil for the base thymine. Different forms of RNA play crucial roles in the translation of DNA's coded instructions for making proteins.

RNA polymerase: an enzyme that unites loose nucleotides to create a strand of messenger RNA, using a segment of DNA as a template.

Sex cell: either a sperm or an egg cell, each containing half the full complement of chromosomes and uniting during conception to form an embryo; also known as a reproductive cell.

Sex chromosomes: in humans, the 23d pair of chromosomes, of which there are two types, X and Y; a female offspring receives one X chromosome from each parent, while a male receives an X from the mother and a Y from the father.

Structural gene: a gene that codes for one of the proteins that make up the structural components of an organism.

Transcription: the process whereby genetic instructions encoded in a segment of DNA are converted into a strand of messenger RNA.

Transfer RNA: a type of RNA that carries amino acids to ribosomes for the purpose of constructing a protein.

Translation: the conversion of genetic information encoded on a strand of messenger RNA into a sequence of amino acids that makes up a specific protein.

Vector: in gene therapy, an agent such as a modified virus that delivers a therapeutic gene to a patient's cells.

Virus: a microorganism that consists of DNA or RNA and various enzymes, all surrounded by a protein coat. Viruses reproduce by invading living cells and using cellular mechanisms to create multiple copies of themselves.

BIBLIOGRAPHY

BOOKS

Alberts, Bruce, et al. *Molecular Biology of the Cell* (2d ed.). New York: Garland, 1989.

Baskin, Yvonne. *The Gene Doctors.* New York: William Morrow, 1984.

Becker, Wayne M. *The World of the Cell.* Menlo Park, Calif.: Benjamin/Cummings, 1986.

Berg, Paul, and Maxine Singer. *Dealing with Genes: The Language of Heredity.* Mill Valley, Calif.: University Science Books, 1992.

Blatt, Robin J. R. *Prenatal Tests.* New York: Vintage Books, 1988.

Bouchard, Thomas J., Jr.:
"A Twice-Told Tale: Twins Reared Apart." In *Personality and Psychopathology* (Vol. 2 of *Thinking Clearly About Psychology: Essays in Honor of Paul Everett Meehl,* ed. by W. Grove and D. Ciccehetti). Minneapolis: University of Minnesota Press, 1991.
"Twins Reared Together and Apart: What They Tell Us about Human Diversity." In *Individuality and Determinism,* ed. by Sidney W. Fox. New York: Plenum, 1984.

Burgess, Jeremy, Michael Marten, and Rosemary Taylor. *Under the Microscope.* Cambridge, Mass.: Cambridge University Press, 1987.

Cassill, Kay. *Twins: Nature's Amazing Mystery.* New York: Atheneum, 1982.

Clayman, Charles B., M.D. (ed.). *The American Medical Association Encyclopedia of Medicine.* New York: Random House, 1989.

Considine, Douglas M., and Glenn D. Considine. *Van Nostrand's Scientific Encyclopedia* (7th ed.). New York: Van Nostrand Reinhold, 1983.

Cummings, Michael R. *Human Heredity.* St. Paul: West Publishing, 1988.

Darnell, James, Harvey Lodish, and David Baltimore. *Molecular Cell Biology* (2d ed.). New York: Scientific American Books, 1990.

De Duve, Christian. *A Guided Tour of the Living Cell* (Vols. 1 and 2). Illus. by Neil O. Hardy. New York: Scientific American Books, 1984.

Family Ties (Successful Parenting series). Alexandria, Va.: Time-Life Books, 1987.

Fast, Julius. *Blueprint for Life: The Story of Modern Genetics.* New York: St. Martin's Press, 1965.

Franck, Irene, and David Brownstone. *The Parent's Desk Reference.* New York: Prentice Hall, 1991.

Freedman, Daniel G. *Human Sociobiology: A Holistic Approach.* New York: The Free Press, 1979.

Gregory, Richard L. *The Oxford Companion to the Mind.* Oxford: Oxford University Press, 1987.

The Incredible Machine. Washington, D.C.: National Geographic Society, 1986.

Judson, Horace Freeland. *The Eighth Day of Creation.* New York: Simon and Schuster, 1980.

Kevles, Daniel J., and Leroy Hood (eds.). *The Code of Codes.* Cambridge, Mass.: Harvard University Press, 1992.

Kunze, Jürgen, M.D., and Irmgard Nippert, M.D. *Genetics and Malformations in Art.* Berlin: Grosse Verlag, 1986.

Lappe, Marc. *Genetic Politics: The Limits of Biological Control.* New York: Simon & Schuster, 1979.

Larrick, James W., M.D., and Kathy L. Burck, M.D. *Gene Therapy.* New York: Elsevier, 1991.

Levine, Arnold J. *Viruses.* New York: Scientific American Library, 1992.

Lewin, Benjamin. *Genes IV.* Oxford: Oxford University Press, 1990.

Lewontin, Richard. *Human Diversity.* New York: Scientific American Books, 1982.

Life Search (Voyage through the Universe series). Alexandria, Va.: Time-Life Books, 1988.

McKusick, Victor A., M.D. *The Morbid Anatomy of the Human Genome.* Baltimore, Md.: Williams & Wilkins, 1986.

Milunsky, Aubrey, M.D. *Choices, Not Chances.* Boston: Little, Brown, 1989.

Mussen, Paul Henry, et al. *Child Development and Personality.* New York: Harper & Row, 1984.

Neubauer, Peter B., M.D., and Alexander Neubauer. *Nature's Thumbprint: The New Genetics of Personality.* Reading, Mass.: Addison-Wesley, 1990.

The New Encyclopaedia Britannica (Vols. 15 and 16). Chicago: Encyclopaedia Britannica, 1989, 1984.

Nichols, Eve K. *Human Gene Therapy.* Cambridge, Mass.: Harvard University Press, 1988.

Nilsson, Lennart. *A Child is Born* (rev. ed.). New York: Dell, 1977.

Nilsson, Lennart, and Jan Lindberg: *Behold Man.* Boston: Little, Brown, 1973. *The Body Victorious.* New York: Delacorte Press, 1987.

1983 Yearbook of Science and the Future. Chicago: Encyclopaedia Britannica, 1982.

Noble, Elizabeth. *Having Twins* (2d ed.). Boston: Houghton Mifflin, 1991.

Papalia, Diane E., and Sally Wendkos Olds. *Human Development* (5th ed.). New York: McGraw-Hill, 1992.

Pierce, Benjamin A. *The Family Genetic Sourcebook.* New York: John Wiley & Sons, 1992.

Raven, Peter H., and George B. Johnson. *Biology* (2d ed.). St. Louis: Times Mirror/Mosby College, 1989.

Schiff, Michel, and Richard Lewontin. *Education and Class: The Irrelevance of IQ Genetic Studies.* Oxford: Clarendon Press, 1986.

Shapiro, Robert. *The Human Blueprint: The Race to Unlock the Secrets of Our Genetic Code.* New York: Bantam Books, 1991.

Singer, Maxine, and Paul Berg. *Genes & Genomes: A Changing Perspective.* Mill Valley, Calif.: University Science Books, 1992.

Singer, Sam. *Human Genetics: An Introduction to the Principles of Heredity* (2d ed.). New York: W. H. Freeman, 1985.

Suzuki, David T., et al. *An Introduction to Genetic Analysis.* New York: W. H. Freeman, 1989.

Suzuki, David T., and Peter Knudtson. *Genethics: The Clash between the New Genetics and Human Values.* Cambridge, Mass.: Harvard University Press, 1989.

Thomas, Lewis. *The Lives of a Cell.* New York: Bantam Books, 1974.

Varmus, Harold, and Robert A. Weinberg. *Genes and the Biology of Cancer.* New York: Scientific American Library, 1993.

Watson, James D., et al. *Molecular Biology of the Gene* (Vols. 1 and 2, 4th ed.). Menlo Park, Calif.: Benjamin/Cummings, 1987.

Watson, James D., et al. *Recombinant DNA* (2d ed.). New York: Scientific American Books, 1992.

Watson, Peter. *Twins: An Uncanny Relationship?* Chicago: Contemporary Books, 1981.

Weaver, Robert F., and Philip W. Hedrick. *Basic Genetics.* Dubuque, Iowa: Wm. C. Brown, 1991.

Whittle, M. J., and J. M. Connor (eds.). *Prenatal Diagnosis in Obstetric Practice.* Oxford: Blackwell Scientific, 1989.

Wills, Christopher. *Exons, Introns and Talking Genes.* New York: HarperCollins, 1991.

A World of Luck (Library of Curious and Unusual Facts series). Alexandria, Va.: Time-Life Books, 1991.

Young, Robert S., Ronald J. Jorgenson, and Steven D. Shapiro. *Odds-R: Chromosomal Book.* San Antonio, Tex.: Grendel, 1985.

PERIODICALS

Aldhous, Peter. "Britain Plans Large-Scale Sequencing Center." *Science*, May 15, 1992.

Allman, William F. "The Story of Life Unfolding." *U.S. News & World Report*, May 22, 1989.

Anderson, Christopher. "Gene Therapy Researcher under Fire over Controversial Cancer Trials." *Nature*, Dec. 3, 1992.

Angier, Natalie:

"A First Step in Putting Genes into Action: Bend the DNA." *New York Times*, Aug. 4, 1992.

"Potent Guardian of Chromosomes: A Muscular Gene Halts Cancer." *New York Times*, Sept. 22, 1992.

Babington, Charles. "U-Md. Cancels Conference on Genetic Link to Crime." *Washington Post*, Sept. 5, 1992.

Barinaga, Marcia. "Signals into Unknown Territory." *Science*, Mar. 27, 1992.

Barry, John M. "Cracking the Code." *Washingtonian*, Feb. 1991.

Beardsley, Tim:

"Clearing the Airways." *Scientific American*, Dec. 1990.

"Making Antisense." *Scientific American*, Feb. 1992.

"Profile: Gene Doctor." *Scientific American*, Aug. 1990.

Begley, Sharon, and Martin Kasindorf. "Identical Twins." *Newsweek*, Dec. 3, 1979.

Benditt, John. "POU! Goes the Homeobox." *Scientific American*, Feb. 1989.

Benne, Rob, and Hans Van der Spek. "L'Editing des Messages Génétiques." *La Recherche*, July/Aug. 1992.

Berreby, David. "Kids, Creoles, and the Coconuts." *Discover*, Apr. 1992.

Bouchard, Thomas J., Jr. "Finding My Twin Brought Happiness Beyond Belief." *Woman*, Sept. 1985.

Bouchard, Thomas J., Jr., et al. "Sources of Human Psychological Differences: The Minnesota Study of Twins Reared Apart." *Science*, Oct. 12, 1990.

Bouchard, Thomas J., Jr., and Matthew McGue. "Genetic and Rearing Environmental Influences on Adult Personality." *Journal of Personality*, Mar. 1990.

Bower, Bruce. "Abusive Inheritance." *Science News*, Nov. 14, 1992.

Brownlee, Shannon. "Immunology's Designer Genes." *U.S. News & World Report*, Oct. 30, 1989.

Caldwell, Mark. "How Does a Single Cell Turn Into a Whole Body?" *Discover*, Nov. 1992.

Carey, John. "The Genetic Age." *Business Week*, May 28, 1990.

Carey, John, and Joan Hamilton. "The Gene Doctors Roll Up Their Sleeves." *Business Week*, Mar. 30, 1992.

Chang, David D., and David A. Clayton. "A Mammalian Mitochondrial RNA Processing Activity Contains Nucleus-Encoded RNA." *Science*, Dec. 13, 1991.

Culliton, Barbara J. "Designing Cells to Deliver Drugs." *Science*, Nov. 10, 1989.

Culver, Kenneth W., et al. "In Vivo Gene Transfer with Retroviral Vector-Producer Cells for Treatment of Experimental Brain Tumors." *Science*, June 12, 1992.

Dajer, Tony:

"Divided Selves." *Discover*, Sept. 1992.

"The Genome Finds Its Henry Ford." *Discover*, Jan. 1993.

De Robertis, Eddy M., Guillermo Oliver, and Christopher V. E. Wright. "Homeobox Genes and the Vertebrate Body Plan." *Scientific American*, July 1990.

Diamond, Jared (ed.). "Curse and Blessing of the Ghetto." *Discover*, Mar. 1991.

Doolittle, Russell F. "Proteins." *Scientific American*, Oct. 1985.

Dorozynski, Alexander. "Gene Mapping the Industrial Way." *Science*, Apr. 24, 1992.

Driscoll, Robert J., Michael G. Youngquist, and John D. Baldeschwieler. "Atomic-Scale Imaging of DNA Using Scanning Tunnelling Microscopy." *Nature*, July 19, 1990.

Ezzell, C.:

"Blood-Vessel Growth Genes Stop Making Sense." *Science News*, Sept. 5, 1992.

"Gene Therapy for Cystic Fibrosis Patients." *Science News*, Dec. 12, 1992.

Farber, Susan. "Telltale Behavior of Twins." *Psychology Today*, Jan. 1981.

Felsenfeld, Gary. "Chromatin as an Essential Part of the Transcriptional Mechanism." *Nature*, Jan. 16, 1992.

"Female Alcoholism Seen Strongly Linked to Genetics." *Washington Post*, Oct. 14, 1992.

Flam, Faye. "Backward Genetics." *Science News*, June 10, 1989.

Fogle, Suzanne. "Pretty BABI: Blastomere Screen Detects CF Gene." *Journal of NIH Research*, June 1992.

"Form-Fitting Genes." *Discover*, Sept. 1987.

Franklin-Barbajosa, Cassandra. "The New Science of Identity." *National Geographic*, May 1992.

Freedman, Daniel G. "Ethnic Differences in Babies." *Human Nature*, Jan. 1979.

Freedman, Daniel G., and Marilyn M. DeBoer. "Biological and Cultural Differences in Early Child Development." *Annual Review of Anthropology*, 1979.

Gehring, Walter J. "The Molecular Basis of Development." *Scientific American*, Oct. 1985.

"Genetics: Influences of Nature, Nurture Debated." *Los Angeles Times*, Apr. 18, 1992.

"The Geography of Genes." *The Economist*, Dec. 16, 1989.

Gibbons, Ann. "Geneticists Trace the DNA Trail of the First Americans." *Science*, Jan. 15, 1993.

Giese, Klaus, Jeffery Cox, and Rudolf Grosschedl. "The HMG Domain of "Lymphoid

Enhancer Factor 1 Bend DNA and Facilitates Assembly of Functional Nucleoprotein Structures." *Cell*, Apr. 3, 1992.

Goldsmith, Marsha F. "Tomorrow's Gene Therapy Suggests Plenteous, Patent Cardiac Vessels." *JAMA*, Dec. 16, 1992.

Goleman, Daniel. "New Storm Brews on Whether Crime Has Roots in Genes." *New York Times*, Sept. 15, 1992.

Gorman, Christine. "A Lethal Legacy." *Time*, May 17, 1993.

Gould, Judy. "What Happens When Three Young Men Find Out They're Triplets? It's Not as Simple as 1-2-3." *People*, Oct. 13, 1980.

Grady, Denise. "Sex Test of Champions." *Discover*, June 1992.

Grunstein, Michael. "Histones as Regulators of Genes." *Scientific American*, Oct. 1992.

Hall, Stephen S. "James Watson and the Search for Biology's 'Holy Grail.'" *Smithsonian*, Feb. 1990.

Haney, Daniel Q. "Born Bashful." *Washington Post*, Mar. 12, 1991.

Hanson, Betsy. "A Bull's-Eye in the Brain." *Discover*, Oct. 1992.

Henig, Robin Marantz. "Dr. Anderson's Gene Machine." *New York Times*, Mar. 31, 1991.

Herman, Robin:

"Gene Therapy to be Tested in Cystic Fibrosis Patients." *Washington Post*, Dec. 4, 1992.

"Scientists to Fight Brain Tumor with Altered Mouse Cells." *Washington Post*, June 9, 1992.

Highfield, Roger. "Gene Injection Used for First Time to Tackle Cancer." *Times* (London), Apr. 14, 1992.

Hoffman, Michelle. "Putting New Muscle into Gene Therapy." *Science*, Dec. 6, 1991.

Holden, Constance:

"The Genetics of Personality." *Science*, Aug. 7, 1987.

"Identical Twins Reared Apart." *Science*, Mar. 21, 1980.

Holden, Constance (ed.). "Another Gene Therapy First." *Science*, June 19, 1992.

Holliday, Robin. "A Different Kind of Inheritance." *Scientific American*, June 1989.

Holloway, Marguerite. "Neural Vector." *Scientific American*, Jan. 1991.

"Homeobox Genes Go Evolutionary." *Science*, Jan. 24, 1992.

Horgan, John. "D_2 or Not D_2." *Scientific American*, Apr. 1992.

Howard-Flanders, Paul. "Inducible Repair of DNA." *Scientific American*, Nov. 1981.

Hunt, Morton. "The Total Gene Screen." *New York Times Magazine*, Jan. 19, 1986.

Hurrell, Mick. "Opening Up the Book of Life." *Times* (London), July 30, 1992.

Jackson, Donald Dale. "Reunion of Identical Twins, Raised Apart, Reveals Some Astonishing Similarities." *Smithsonian*, Oct. 1980.

Jaroff, Leon:

"Giant Step for Gene Therapy." *Time*, Sept. 24, 1990.

"Happy Birthday Double Helix." *Time*, Mar. 15, 1993.

Jones, Steve. "The Language of Genes: Change or Decay." *The Independent*, Nov. 21, 1991.

Kanigel, Robert. "Genome." *New York Times Magazine*, Dec. 13, 1987.

Kiester, Edwin, Jr. "A Bug in the System." *Discover*, Feb. 1991.

Kluger, Jeffrey. "Body Doubles." *Omni*, Aug. 1987.

Kolata, Gina:

"Fetal Hemoglobin Genes Turned On in Adults." *Science*, Dec. 24, 1982.

"From Fly to Man, Cells Obey Same Signal." *New York Times*, Jan. 5, 1993.

Lang, John S. "How Genes Shape Personality." *U.S. News & World Report*, Apr. 13, 1987.

Lantiéri, Marie-Françoise. "À la Découverte du Génome Humain." *Science & Vie*, Feb. 1988.

Lemonick, Michael D. "Genetic Tests under Fire." *Time*, Feb. 24, 1992.

Leo, John:

"Exploring the Traits of Twins." *Time*, Jan. 12, 1987.

"Genetic Advances, Ethical Risks." *U.S. News & World Report*, Sept. 25, 1989.

Lewis, Ricki. "DNA Fingerprints: Witness for the Prosecution." *Discover*, June 1988.

Lowenstein, Jerold M. "Whose Genome Is It, Anyway?" *Discover*, May 1992.

McAuliffe, Kathleen. "Born to Believe: Your Values about God, Home, and Country May Be Influenced by Your Genes." *Omni*, Oct. 1992.

McKnight, Steven Lanier. "Molecular Zippers in Gene Regulation." *Scientific American*, Apr. 1991.

Maranto, Gina, and Shannon Brownlee. "Why Sex?" *Discover*, Feb. 1984.

Marx, Jean L.:

"The Continuing Saga of 'Homeo-Madness.'" *Science*, Apr. 11, 1986.

"DNA Fingerprinting Takes the Witness Stand." *Science*, June 1988.

Merz, Beverly. "Designer Genes." *American Health*, Mar. 1993.

Moffat, Anne Simon. "Making Sense of Antisense." *Science*, Aug. 2, 1991.

Montgomery, Geoffrey. "The Ultimate Medicine." *Discover*, Mar. 1990.

Mullis, Kary B. "The Unusual Origin of the Polymerase Chain Reaction." *Scientific American*, Apr. 1990.

Nash, J. Madeleine. "Tracking Down the Killer Genes." *Time*, Sept. 17, 1990.

"Now Robots Have Joined the War against Disease." *The Independent*, Nov. 9, 1992.

Okie, Susan. "Genes Play Heavy Role in Weight Gain." *Washington Post*, May 24, 1990.

"100 Trillion Ants Drop Acid." *Discover*,

Sept. 1987.
Palca, Joseph:
"Britain Plans Large-Scale Sequencing Center." *Science*, May 15, 1992.
"The Genome Project: Life after Watson." *Science*, May 15, 1992.
"The Other Human Genome." *Science*, Sept. 7, 1990.
"Peeking at Twins." *The Economist*, Nov. 3, 1990.
Perricaudet, Michel, Leslie Stratford-Perricaudet, and Pascale Briand. "Gene Therapy by Adenovirus." *Le Recherche*, Apr. 1992.
Perutz, M. F. "The Hemoglobin Molecule." *Scientific American*, Nov. 1964.
Price, James O., et al. "Prenatal Diagnosis with Fetal Cells Isolated from Maternal Blood by Multiparameter Flow Cytometry." *American Journal of Obstetrics and Gynecology* (part 1), Dec. 1991.
Ptashne, Mark. "How Gene Activators Work." *Scientific American*, Jan. 1989.
Radetsky, Peter. "The Roots of Cancer." *Discover*, May 1991.
Randall, Teri. "First Gene Therapy for Inherited Hypercholesterolemia a Partial Success." *JAMA*, Feb. 17, 1993.
"Reading the Human Blueprint." *U.S. News & World Report*. Dec. 28, 1987/Jan. 4, 1988.
Richards, Frederic M. "The Protein Folding Problem." *Scientific American*, Jan. 1991.
Ried, Thomas, et al. "Simultaneous Visualization of Seven Different DNA Probes by *In Situ* Hybridization Using Combinatorial Fluorescence and Digital Imaging Microscopy." *Proceedings of the National Academy of Science*, Feb. 1992.
Roberts, Leslie. "Flap Arises over Genetic Map." *Science*, Nov. 6, 1987.
Rosen, Claire Mead. "The Eerie World of Reunited Twins." *Discover*, Sept. 1987.
Segal, Nancy L. "The Nature vs. Nurture Laboratory." *Twins*, July/Aug. 1984.
Stone, Richard. "Molecular 'Surgery' for Brain Tumors." *Science*, June 1992.
"Study Discounts Genetics in Most Alcoholism." *Washington Post*, Jan. 20, 1992.
"Switched-On Genes." *Newsweek*, Dec. 20, 1982.
Thompson, Dick. "The Most Hated Man in Science." *Time*, Dec. 4, 1989.
Thompson, Larry:
"At Age 2, Gene Therapy Enters a Growth Phase." *Science*, Oct. 30, 1992.
"Stem-Cell Gene Therapy Moves Toward Approval." *Science*, Feb. 28, 1992.
"The Tiniest Transplants." *The Economist*, Apr. 25, 1992.
Trotter, Robert J. "You've Come a Long Way, Baby." *Psychology Today*, May 1987.
Verma, Inder M. "Gene Therapy." *Scientific American*, Nov. 1990.
Vogel, Shawna:
"The Case of the Unraveling DNA." *Discover*, Jan. 1990.
"The Shape of Proteins." *Discover*, Oct. 1988.
Watson, Traci. "Gene Therapy Research Is Spread across Many Disciplines and Institutes at NIH." *Nature*, Sept. 17, 1992.
Weaver, Robert F. "Changing Life's Genetic Blueprint." *National Geographic*, Dec. 1984.
Weinberg, Robert A. "A Molecular Basis of Cancer." *Scientific American*, Nov. 1983.
Weintraub, Harold M. "Antisense RNA and DNA." *Scientific American*, Jan. 1990.
Weiss, R. "First Human Gene-Therapy Test Begun." *Science News*, Sept. 22, 1990.
Wickramasinghe, H. Kumar. "Scanned-Probe Microscopes." *Scientific American*, Oct. 1989.
Wright, Robert. "Achilles' Helix." *The New Republic*, July 9/16, 1990.

OTHER SOURCES

Bouchard, Thomas J., Jr., Nancy L. Segal, and David T. Lykken. "Genetic and Environmental Influences on Special Mental Abilities in a Sample of Twins Reared Apart." Rome: The Mendel Institute, 1990.
DNA *Identity Testing*. Stamford, Conn.: Lifecodes Corporation, 1991.
The Human Genome Project: New Tools for Tomorrow's Health. Bethesda, Md.: National Institutes of Health, Sept. 1992.
Kanigel, Robert. "How Does a Fertilized Egg Turn Into a Fly, a Chicken, You, or Me?" *From Egg to Adult*. Report. Bethesda, Md.: The Howard Hughes Medical Institute, May 1992.
"Note to Reporters and Editors." News release. Bethesda, Md.: National Institutes of Health, Oct. 8, 1991.
Pines, Maya. "Learning from the Worm." *From Egg to Adult*. Report. Bethesda, Md.: The Howard Hughes Medical Institute, May 1992.
Radetsky, Peter:
"Discovering the Body Plan." *From Egg to Adult*. Report. Bethesda, Md.: The Howard Hughes Medical Institute, May 1992.
"The Homeobox: Something Very Precious that We Share with Flies." *From Egg to Adult*. Report. Bethesda, Md.: The Howard Hughes Medical Institute, May 1992.
Rosenberg, Steven A. "The David A. Karnofsky Memorial Lecture: Immunotherapy and Gene Therapy of Cancer." American Society of Clinical Oncology 27th Annual Meeting, Houston, Tex., May 20, 1991.
U.S. Congress, Office of Technology Assessment. *Mapping Our Genes—The Genome Projects: How Big, How Fast?* OTA-BA-373. Washington, D.C.: U.S. Government Printing Office, Apr. 1988.
U.S. Department of Energy. *Human Genome Program—Primer on Molecular Genetics*. Washington, D.C.: Office of Energy Research, June 1992.

INDEX

Numerals in bold refer to illustrations or illustrated text.

A

B

C

ACKNOWLEDGMENTS

The editors of *Blueprint for Life* wish to thank these individuals for their valuable contributions:

Vanessa M. Barnabei, George Washington University, Washington, D.C.; David Bigbee, Federal Bureau of Investigation, Washington, D.C.; Thomas J. Bouchard, Jr., Minnesota Center for Twin and Adoption Research, Minneapolis; Patricia Bray-Ward, Yale University, New Haven, Connecticut; Karen Camille Cone, University of Missouri, Columbia; Kim Cornejo, Howard Hughes Medical Institute, Chevy Chase, Maryland; Jackie Cossman, Applied Biosystems, Foster City, California; Alan Cowell, Rome; Maurizio Cutolo, Genoa; Philippe Dacla, CNRI, Paris; Paolo Durand, Rome; Sharon Durham, National Institutes of Health, Bethesda, Maryland; Nina V. Fedoroff, Carnegie Institution of Washington, Baltimore; Gary Felsenfeld, National Institutes of Health, Bethesda, Maryland; Jill S. Fonda-Allen, Children's National Medical Center, Washington, D.C.; Hans Gelderblom, Robert-Koch-Institute, Berlin; Michael Gibson, Applied Biosystems, Foster City, California; Mark R. Hughes, Baylor College of Medicine, Houston, Texas; Lynn Jorde, University of Utah, Salt Lake City; Margaret Keyes, Minnesota Center for Twin and Adoption Research, Minneapolis; Troels Kjaer, National Institute of Mental Health, Bethesda, Maryland; Victor McKusick, The Johns Hopkins University, Baltimore; Christiane Nüsslein-Volhard, MPI für Entwicklungsbiologie, Tübingen; Michael E. Pique, The Scripps Research Institute, La Jolla, California; Tom Reynolds, National Institutes of Health, Bethesda, Maryland; Michael B. Wiant, Illinois State Museum, Springfield.

PICTURE CREDITS

The sources for the illustrations for this book are listed below. Credits from left to right are separated by semicolons; credits from top to bottom are separated by dashes.

Cover: KHS/Okapia. **7:** Art by Alfred T. Kamajian—Drs. T. Reid & D. Ward/Peter Arnold, Inc.—© 1993 Bob Sacha. **8, 9:** Lennart Nilsson, *The Body Victorious*, Dell Publishing Co., New York, 1987. **11:** Enrico Ferorelli. **12, 13:** Phil Schofield/Allstock. **14:** Richard G. Hatton, *The Craftsman's Plant-Book*, 1909 (Dover Edition, New York, 1960)—chart by Fatima Taylor. **16:** © RSI/Allstock. **18, 19:** Art by Will Williams of Wood, Ronsaville, Harlin, Inc. **20, 21:** Rolf Konow/Shooting Star. **23:** Art by Alfred T. Kamajian. **24, 25:** Art by Alfred T. Kamajian; (top right) Lennart Nilsson, *A Child is Born*, Dell Publishing Co., New York, 1986—(bottom right) Lennart Nilsson, *Behold Man*, Little, Brown, Boston, 1973, (insets) Biophoto Associates/Science Source/Photo Researchers (3). **26:** Dr. Per Sundström/CNRI—© Biophoto Associates/Science Source/Photo Researchers (2)—Charlotte Fullerton (2). **28, 29:** Inset art by Tipy Taylor; Manfred P. Kage/Okapia (2)—art by Alfred T. Kamajian; Georgetown University Medical Center. **31:** Dr. Jürgen Kunze, Kinderklinik und Institut für Hümangenetik, Freie Universität, Berlin. **32:** Art by Tipy Taylor. **33:** Art by Fred Holz. **34:** Peter Freed. **36, 37:** The Mansell Collection; chart by John Drummond; The Royal Archives, Windsor Castle © Her Majesty Queen Elizabeth II (2). **39:** Michael Pique/Research Institute of Scripps Clinic. **40, 41:** Art by Will Williams and Rob Wood of Wood, Ronsaville, Harlin, Inc. **42-49:** Art by Rob Wood of Wood, Ronsaville, Harlin, Inc. **50, 51:** Will & Deni McIntyre/Photo Researchers. **52:** Driscoll, Youngquist & Baldeschwieler, Caltech/Science Photo Library/Photo Researchers. **54, 55:** Petit Format/Nestle/Science Source/Photo Researchers. **56:** CNRI/Science Photo Library/Photo Researchers—CNRI/Science Source/Photo Researchers. **57:** CNRI/Phototake, New York—Prof. P. Motta/Department of Anatomy/University La Sapienza, Rome/Science Photo Library/Photo Researchers. **59:** David Scharf/Peter Arnold, Inc. **60:** Dr. Christiane Nüsslein-Volhard, Tübingen; Steve Langeland, Steve Paddock, Sherwin Attai, and Sean Carroll/Howard Hughes Medical Institute Research Laboratories. **61:** Steve Langeland, Steve Paddock, Sherwin Attai, and Sean Carroll/Howard Hughes Medical Institute Research Laboratories (2); © 1992 *Discover* Magazine. **63:** Alan Oddie/PhotoEdit. **64, 65:** Lennart Nilsson, *The Body Victorious*, Dell Publishing Co., New York, 1987, (inset) David A. Clayton/Fran Heyl & Associates. **67:** Secchi-Lecaque/Roussel-UCLAF/CNRI/Science Photo Library/Photo Researchers. **68:** J. C. Revy, CNRI. **69:** Bill Longcore/Photo Researchers. **71:** Lee D. Simon/Photo Researchers—Dr. Hans Gelderblom, Robert Koch-Institut, Berlin. **72, 73:** Art by Alfred T. Kamajian. **75:** Nik Kleinberg/Picture Group. **77:** Art by Steven Bauer—courtesy of The Illinois State Museum. **78, 79:** Art by Steven Bauer; Lifecodes Corporation. **80:** © Boehringer Ingelheim International GmbH, photo Lennart Nilsson. **85:** Drs. T. Ried & D. Ward/Peter Arnold, Inc. **86:** Courtesy National Center for Human Genome Research, Bethesda, Maryland. **87:** Courtesy *The New Republic*. **88, 89:** © Jean-Claude Revy/Phototake, New York. **90-95:** Art by Karen Barnes of Wood, Ronsaville, Harlin, Inc. **96, 97:** Photographer Kay Chernush, for Howard Hughes Medical Institute. **98-105:** Art by Stephen R. Wagner. **106, 107:** © David Teplica, M.D., M.F.A. **109:** © 1993 Robert Mankoff and The Cartoon Bank, Inc. **111:** Robert Halmi. **113:** © David Teplica, M.D., M.F.A. **114:** Doris Brenner. **116, 117:** The Granger Collection, New York. **118:** London Express/Editorial Enterprises. **119:** © 1987 Bob Sacha. **120:** Robert Burroughs. **121:** © 1987 Bob Sacha. **122:** UPI/Bettmann. **123:** © 1987 Jim Coit. **125:** Courtesy Dr. Monte S. Buchsbaum. **128-131:** Background Dan McCoy/Rainbow. **128, 129:** Culver Pictures, Inc.—Manfred P. Kage/Okapia—courtesy of The Rockefeller Archive Center—Photo Researchers. **130, 131:** H. R. Kobel/Université de Genève—Solo Syndication/Sipa Press—Dr. Ralph Brinster—© M. Beret/RHPHU/Photo Researchers—J. C. Revy/CNRI.